Transparency, Form and Function

Fermenter manufacturing – Art

Credits

Design, typesetting: Heinz Wild Design, Zürich
Essay: Ludwig Hasler, Zollikon
Art: Rudolf Butz, Solothurn
Technical illustrations: Patrick Graf, Zürich
Photography: Herbert Thür, Basel
Technical descriptions: Paolo Sarti, Bioengineering AG, Wald
Translations: Ueli Hepp, Wald; Erich Brandenberger, Zollikon
Printing: Fotorotar AG, Egg
Binding: Buchbinderei Burkardt AG, Mönchaltorf

ISBN 3-905409-10-0

Man, what a beautiful maschine!

Notes on the relation between machines and aesthetics

Ludwig Hasler

Now, what's Bioengineering up to this time? The company invents remarkably intelligent machines. And sells them successfully around the globe. Shouldn't that be enough? Why is it now elevating these machines into the realm of art? Does it want to ennoble its products by aestheticizing them? To rid them of the commonplace flavor of utility?

In the hands of artist Rudolf Butz, the sterile components lose their sterility, their clarity of purpose melts into ambiguity. Transformed into art, they become self-sufficient, self-sustaining universes in their own right, sometimes waving the alchemist's wand, sometimes contemplatively marveling at their own creative forces. Each one of them has its unique colorful, dynamic inner life, detached from all human intervention. Homo Faber has vanished from the scene, the machine has gained independence, requiring no inventor, no operator, no user. The machine creates itself.

What is Bioengineering trying to tell us? That its machines are not prosaic pumping devices but

wonderfully autonomous creations? In keeping with the motto of the sorcerer's apprentice: Man proposes, Machine disposes? How should we interpret the attitude behind all this? Is it the Faustian urge for presumptuousness – or is it the modesty of the craftsman for whom the result matters more than all else? In either case, the machines take center stage. And aestheticized by Rudolf Butz, they put on a phenomenal show.

The choreography of this book provides a first answer. It contrasts Butz's images with the scientifically accurate drawings by Patrik Graf. Their subject is the same: fermentation technology products from the town of Wald ZH. However, the two approaches could hardly be more disparate. With creative panache, Butz takes hold of the machines, while Graf dissects them with the meticulous three-dimensional vision of the surgeon. The outcome is an equally fascinating revelation of their inner workings. Where Butz discovers something akin to the soul of the machine, Graf exposes its material functionality, affords a glance at the maze of cables, articulations and valves. The two artists complement each other like the psychologist and the neuroscientist. The one searches for meaning, the other for function. Butz generates interpretations, Graf, transparence.

For laypeople like me, this transparent technical inner life is simply a feast for the eye. Engineers will see it differently. For them, the drawings eliminate the effort required in industrial espionage. Quite obviously, Bioengineering can afford to be unconcerned. Either its designers have already devised much more sophisticated machines or the company is willing to take the risk of intellectual property theft because its focus is elsewhere. If so, where?

I am only marginally acquainted with the company. But even from the margin, it is obvious: Bioengineering is serious about the notion that art and technology, culture and work can cross-fertilize each other. While most companies remain firm in their conventional conviction that culture is something like the privately measured dash of emotional warmth in the cold pursuit of profitability, Bioengineering has for years been infusing culture into its daily business. It publishes proprietary literary works, and it invites its employees to lectures, readings and concerts at regular intervals – during working hours.

This might not be purely altruistic. After all, employees tend to be much more attentive and inventive when, besides focusing their minds on the work at hand, they have also sharpened their senses. In modern companies, people do not merely do their jobs and follow the instructions of their superiors. Instead, as their own entrepreneurs, they keep reinventing their work. To do this, their senses must be keen, wide awake, eagerly innovative. It is these senses that want to be aroused, stimulated, inspired. And art can accomplish this better than any other agent.

Friedrich Schiller called it *The Aesthetic Education of Mankind.* Education is no longer a tool any executive board can wield, but the project in the small Swiss town of Wald points in a similar direction: human beings, more creatures of the senses than beasts of reason, need the stimulation and inspiration of the arts if their head is to keep alive its connection to imagination and creative intelligence. Aesthetic education can be a recipe for success in a company that thrives on innovation. Technical intelligence and economic rationality may deliver answers to questions of How, but to answer the decisive What question, engineers need more than professional competence: a vivid sense of feasibility is essential. However, this is not part of nature's dowry; it wants to be tickled, nurtured, refined. Nothing promotes it as productively as the arts, these lovers of the incalculable, these eroticists of the unpredictable.

This book represents a peak in Bioengineering's aesthetic education project. It transforms the company's finished products back into *potentialities*. The very artifacts they have themselves produced are meant to fuel the employees' imagination. This may be construed as slyness (employees are being catered to from their own cellar), but equally well as recognition (employees perceive the fruit of their labor as fantastic objects). Whatever the case may be, the project has a secret message: surround yourself with aesthetically pleasing things and you will discover the extent of your own sensory faculties.

The question remains: what are aesthetically pleasing things? The tradition of aesthetics used to distinguish three kinds of beauty: natural beauty, artistic beauty, and functional beauty. There is no need to dwell on the truism that *nature* may be perceived as beautiful. All you need to do is hike down from Piz Nair into the valley below with your eyes open. People who are not moved by the silent majesty of those mountains are sensual clods. It is equally self-evident that *art* can be beautiful. Whoever can listen to Giacomo Puccini's opera "La Bohème" without being touched by Mimi's dying music has been emotionally dead for a long time. But what makes functional objects beautiful?

I would like to share a few thoughts on this with you.

Machines and beauty. As trivial as it may sound: what are machines for if not to be used? Years ago, as an editor-in-chief, I used to stand in awe in front of the rotary press, this exorbitantly expensive high-tech printing machine that spits out 35,000 fat newspaper copies an hour, like an army of invisible workers slapping text and images on paper and daubing the right ink in the right place at the right time. A marvel of technology – and a delight to behold. It rumbles and undulates, the paper rushes over the rollers and cylinders, and in the end is accurately folded, stacked and addressed.

How I admired it all ... until things went awry. A slightly wrong shade of ink, too much magenta, or a complete standstill, a torn web – and the technical marvel seemed to have morphed into one of those absurd Tinguely machines. Like a piece of *l'art pour l'art.* Art may very well serve its own purpose, but a machine is definitely a means to an end. If it fails, no art can put it back on track. On the contrary: in such cases, all artistry becomes merely annoying. The other day I noticed that my razor was losing its edge. An hour later, there was a commercial on TV that said, "This shave will make any woman swoon," or something to that effect. I went and bought the competitor's product.

Our world of goods is full of beautiful things. Bang & Olufson, for example, the entertainment systems, superb design. Unfortunately, I must say as a music buff, without sensuality, without feeling for the subtleties of tone. Everything pours from the speakers in level uniformity. But can a machine have a soul? Certainly, for Rudolf Butz. Perhaps even for a scientific artist like Patrik Graf. But for the ordinary user?

Or is functional beauty merely eyewash, cosmetics, design? As early as 1930, the futurists proclaimed that a racing car was more beautiful than the Venus of Milo. That was a provocation, today it is a platitude. The borderlines between beautiful bodies, beautiful art, and beautiful industrial products are dissolving. The same glamour illuminates bodies, art, machines. Expensive cars look incredibly sensuous, beautiful women incomparably precious. Is beauty falling victim to commercial aesthetics?

Let us see. The question is: what makes useful things beautiful? A lighter, a watch, a fountain pen, a power drill, a dress, a car, a sterile component? Does the "beauty" of everyday objects depend on their functionality – or on their design? Le Corbusier, for instance, declared an object to be beautiful if it was "sans ambiguïté," unambiguous. If we accept this view, we will have to admit that the real sterile components from Wald are more beautiful than those artistically transformed by Rudolf Butz.

I happen to be sitting on a Corbusier chair. Without doubt a sensuous pleasure. Nonetheless, I have an ambiguous relationship to functional beauty. A personal anecdote may illustrate what I mean. I once bought an old 1964 Jaguar. It looked "beautiful" to me. Why? First, it was a feast for the eye: elegant lines, noble materials. Second, it accelerated awesomely and seemed to float above the roadway, not silently, mind you, but with this slight guttural hum that sounds like music. And finally, there were creature comforts galore: on getting in, resting in the seat, steering, and enjoying the view.

Again: why did I consider it "beautiful"? Not because it was pleasant to look at. Not because it worked well. But because – and this is the essential point – because it intensified ad infinitum my desire to use it. Such a beautiful car is wonderful to use, but even more importantly: it declares utility to be of minor importance. It elevates driving, normally a means to an end, to an end in itself. As soon as I was in the driver's seat, I felt indescribably human. Leather beneath me, mahogany in front of me, the bar behind my back: this car was more than a vehicle. It pampered me, transformed me, ennobled me, animated my senses, refined my penchant for speed, polished the gentleman in me. In other cars, I am just another motorist with only one thing on his mind: what time will I arrive where I am going? In my vintage Jaguar, I became a human being, my senses soared, my thoughts sprouted wings. It was great not to have arrived yet, I was enjoying my time, I was free.

Free until, a few months later, I had to top up two liters of oil each time I stopped at a service station. The car was a pile of rust. Suddenly, I did not find it beautiful any longer, even though it went on looking impressive. This is the difference between functional and artistic beauty. Pure art never needs to function, the functional object always has to work. Nonetheless, useful things are not restricted to their utility. Their aesthetic appeal adds to their functionality. They are beautiful because they arouse and highten the sensuous pleasure we experience as we use them.

Every era defines functional beauty in its own terms. In aristocratic societies, things were considered beautiful if they were visibly costly and artistically crafted. These attributes made them demonstrable symbols of status and power. Bourgeois society adopted this aesthetics of imposition but domesticated, introverted and jovialized it. The cooler the progress of industrialization in the world outside, the warmer the interiors had to be. Now, things were "beautiful" – from coffee cups to curtains – if their function was obliterated beyond recognition by expensive trumpery. The egalitarian society did not succeed in establishing itself aesthetically until the 20th century. The functionalism of modernity gradually eradicated the upper-class tinsel tic and revolutionized the design of useful things. Louis Sullivan coined the apt phrase "form follows function" to describe the underlying principle. This was the end of artistic hypertrophy that had burdened everyday objects. A door knob is a door knob, not a prestigious contraption, and the most beautiful doorknob is the one that feels most comfortable to the hand. Bauhaus minimalism liberated things from the threat of suffocating beneath decorative frippery. It set people free from the musty overelaboration of functional objects. Since then ornamentation has been declared a violation of the freedom of things and human beings.

This, more or less, is still the going doctrine. Remarkably unscathed, it has survived the relatively short postmodern era during which the randomness of form was celebrated with capricious seriousness. "Form follows function" is no doubt an aesthetic creed that offers respiratory relief. Who on earth would want their lifestyle dictated by a utilitarian world embellished with stylistic exuberance? A few odd individuals with a shaky sense of taste, perhaps. "I do not expect the city to humor me," Karl Kraus said. "I am capable of humoring myself." Individuals of style and stature feel patronized if superfluous decoration surrounds them.

The aesthetic allergy reaction against anything superfluous befits modern individuals who want to forge their own lifestyles, although the insistence with which it often articulates itself renders it somewhat suspicious. It was acceptable as a reaction to the aesthetic luxuriance of the bourgeoisie. But today, formal asceticism resembles the petty moralizing of puritans. Despite our unobjectionable aversion to luxuriating décor, we do not perceive an object as "beautiful" unless it radiates a certain charm. As different as individual concepts of beauty may be, a thing must have some facet worth looking at. Pure functionality offers nothing to the sense of beauty, but much to the urge for manipulation. In order to legitimize the talk of functional beauty, therefore, the functional form must be accompanied by something more, something that attracts the eye, invites it to stay, stimulates it. Some sensual surplus beyond the logic of purpose.

Let us cast a last glance at the sterile components. What is there to see? In the drawings, they reveal the secrets of their technical construction. But at the very moment of this revelation, the secret vanishes, instantly. The rational mind understands how the stirrer works: aha, this is how it stirs. That's it. Or, is it? After all, the drawing does not show the stirrer as it stirs. It outlines the idea of a stirrer, not its state of completion. What we see is the stirrer in its nascent state – or, to be more precise: rather than the actual stirrer itself, we see its invention. We are not looking at the stirrer, we are peeping into the engineer's mind. Just as we see more than a hand or a face when we look at a sketch by Leonardo da Vinci: we participate in the artist's act of creation.

What is beautiful is never the finished product, no matter how easily we can handle it. Beauty is what transforms us – as beholders or users – from creatures of fact into perceivers of potentiality. Whether in the form of Leonardo da Vinci's sketch, Tinguely's Heureka machine, Rudolf Butz's photographs or Patrik Graf's drawings of sterile components: art always suspends our everyday frame of mind, fetches objects back into a state of transformability, and turns us into creators. For a brief moment, we slip into the artist's creative impulse, into the engineer's inventive brain center, and discover what we could be if we dared to trust our own designing powers.

I and the world: a cycle of potentiality – a game and a most serious endeavor at once. This is the crucible of human autonomy, the secret of any vital society. And therefore the recipe for economic prosperity.

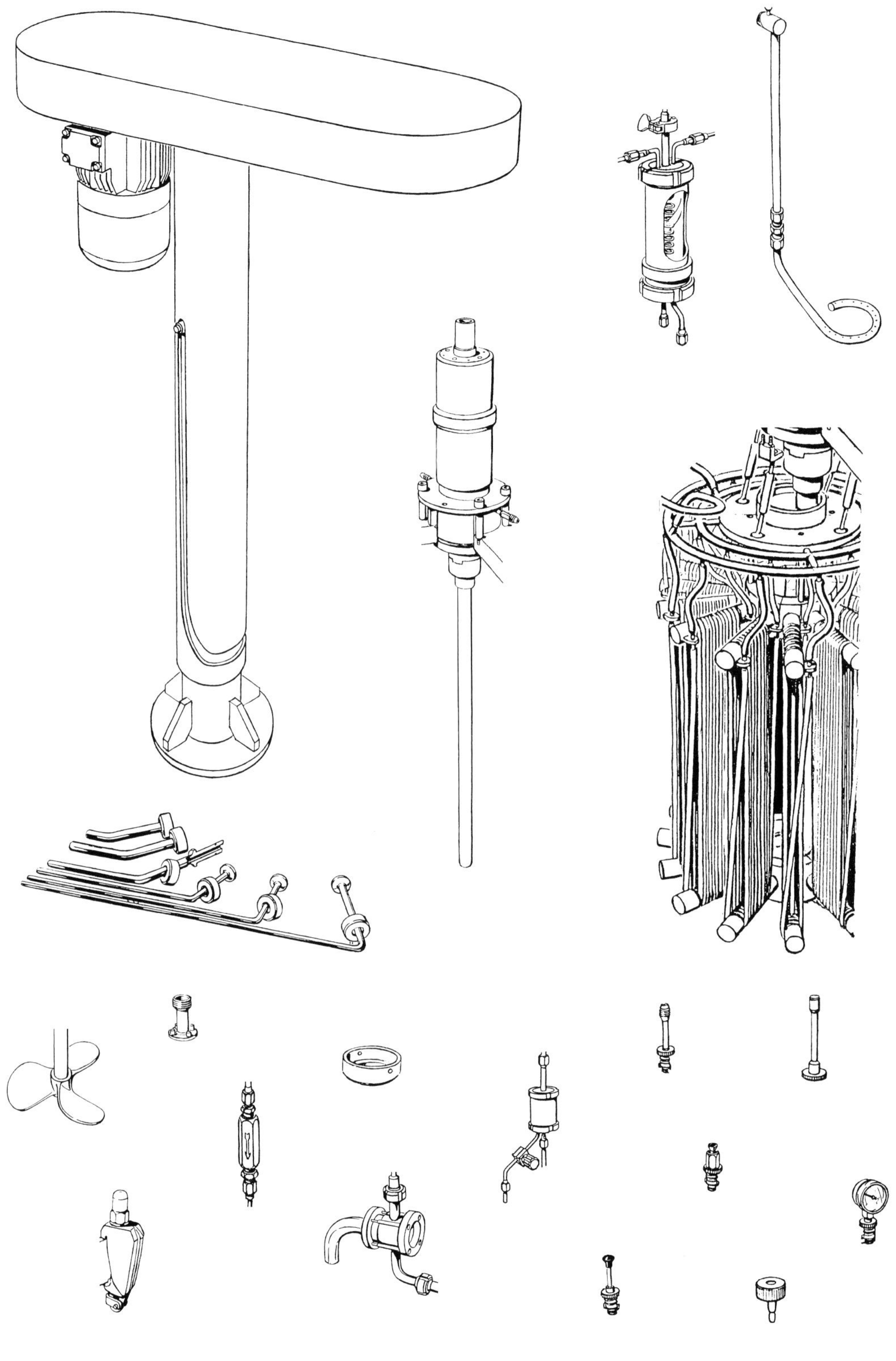

At the beginning of a project, ideas are often only approximate: no precise vessel geometries, only a few characteristics of the medium and the stirring tasks, provisional production environments that have yet to be modified, etc., general conditions and criteria that eventually determine the final configuration of the fermentation system. As the project progresses, the requirements of the specific process become more and more tangible, and step by step, each detail, each component is defined – from the steam trap to the complete reactor with all its elements and connections. The following pages present an overview of the considerations that are involved in the construction of fermenters.

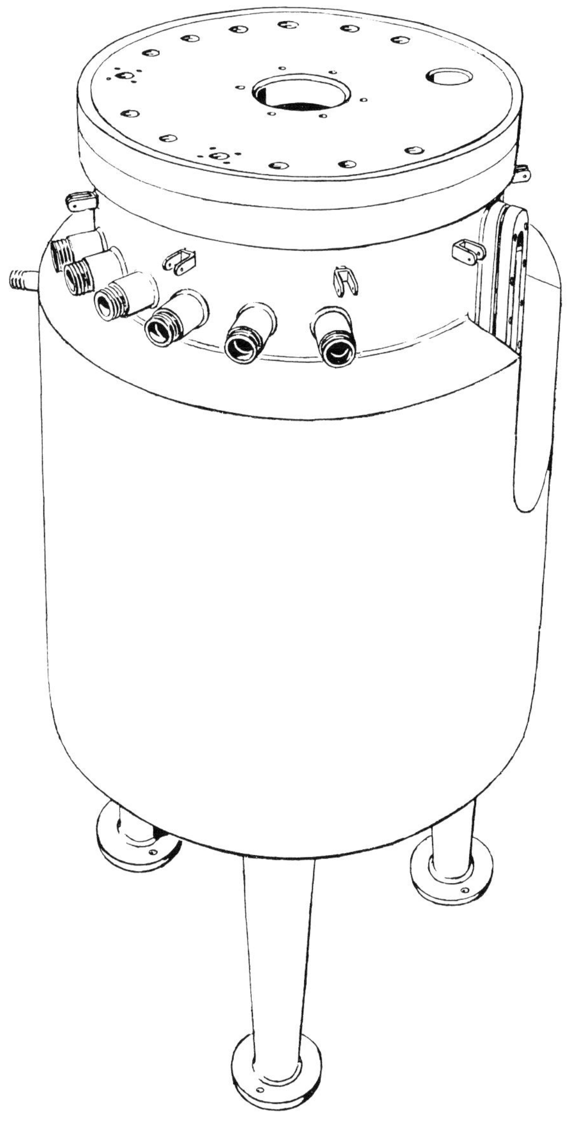

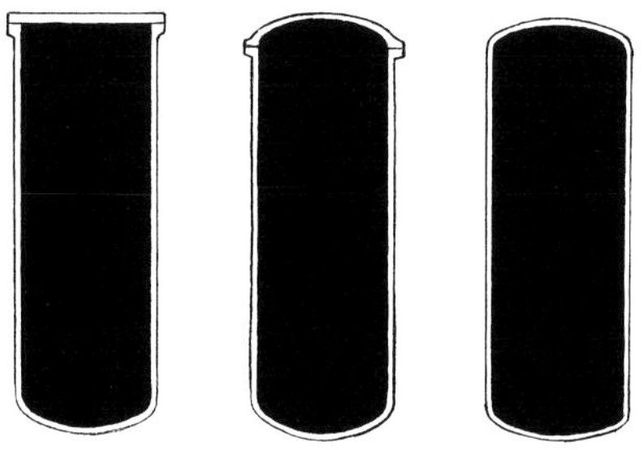

Tall, slender shapes:
Bacterial cultures, etc.

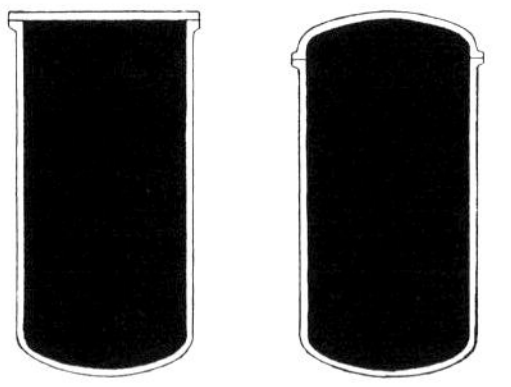

Balanced shapes:
Bacterial / cell cultures,
feed tanks

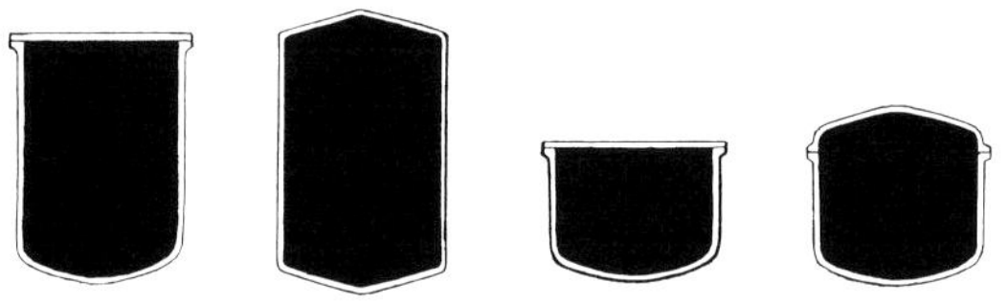

Wide, deep shapes:
Cell cultures, media,
CIP, storage tanks

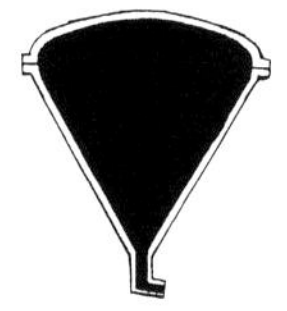

Tapered shapes:
Filtration

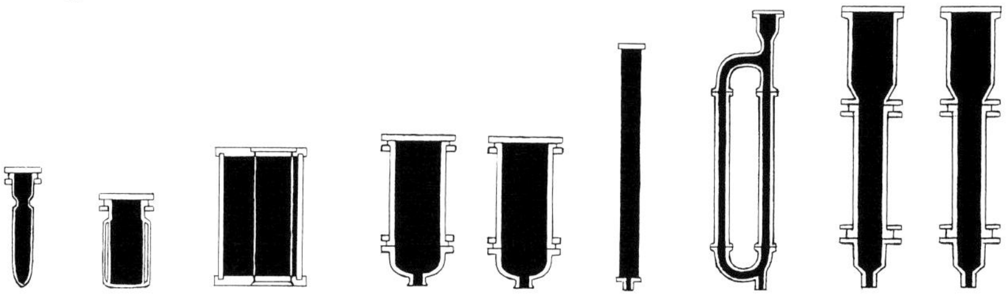

Special shapes

Vessel shapes are determined by their function. In principle, vessels are subdivided into pressurized and non-pressurized vessels. As a rule, non-pressurized vessels are used for the storage of media and do not have to meet very stringent criteria. In most cases, they are thin-walled and wide. In the production of microbial cultures, for instance, narrow and tall vessel shapes are used. Depending on diameter and size, they are designed with either flat or domed bottoms. Narrow vessel shapes allow a high power input resulting in good mixing, efficient aeration, and high heat transmission. For cell cultures requiring gentle treatment, wider shapes are employed. Apart from these classical designs, many different special designs have been developed.

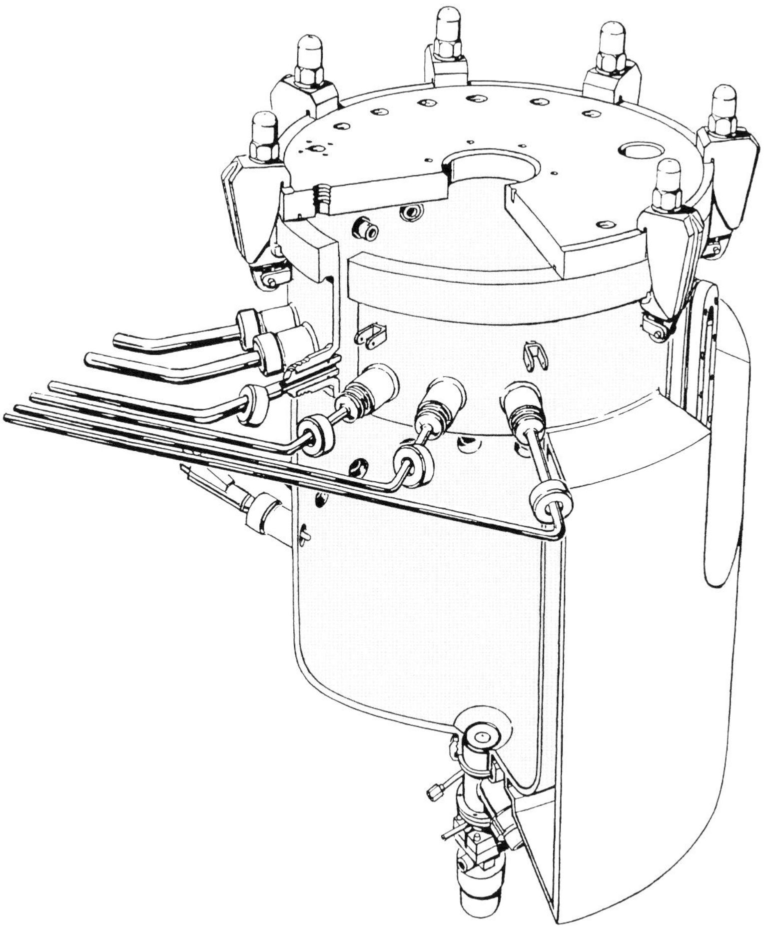

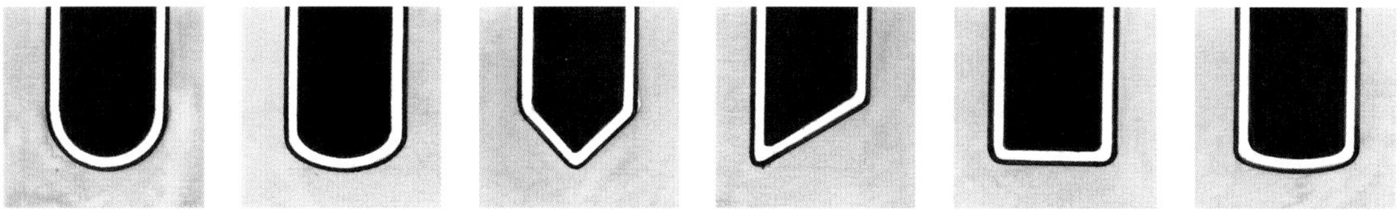

a b c d e f

a Spherical bottom
b Torospherical bottom
c Conical bottom
d Slanted bottom
e Flat bottom
f Plate bottom

The shape of the vessel bottom depends on a number of criteria. The most important are the function of the bottom in the process (e.g. mixing parameters in the vessel, idling properties) and the operating conditions (pressure) to which the bottom is exposed. The principal rule is: The greater the service pressure, the better suited are round-shaped bottoms because the wall thickness, given equal operating conditions, may be lower. Non-pressurized vessels have thinner walls and therefore offer a greater variety of design options.

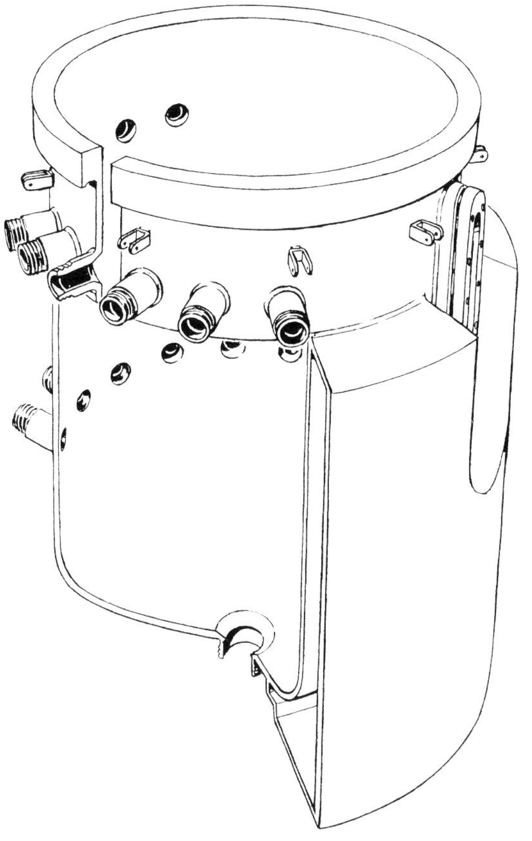

Most bioreactors have torospherical bottoms.

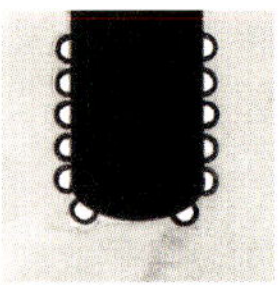

Half-pipe coil

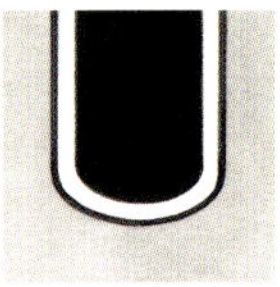

Double jacket

Half-pipe coil combined with double jacket

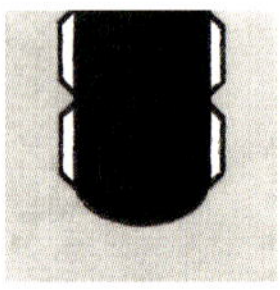

Dimple jacket

Internal pipe coil

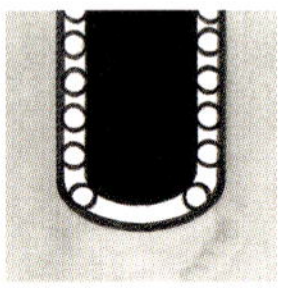

Double jacket with internal pipe coil

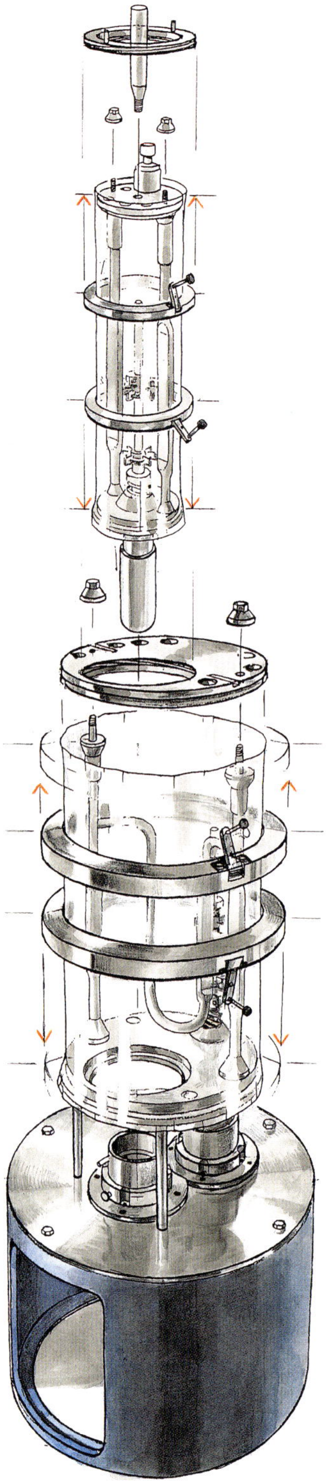

Diagram of an interior heating pipe with baffle function

The figures above show cross sections of customary heating / cooling surfaces. A basic distinction is made between heating surfaces on the outer wall of the reactor and interior exchange surfaces. Since the exterior surface is geometrically limited, a lack of exchange surface may be compensated by creating additional surface inside the reactor, e.g. by pipe bundles. The drawback of such solutions is the risk of contamination of the medium with the heating / cooling agent (through leaks or other defects). Moreover, interior solutions affect the hygienic design of the reactor and therefore require adequate space. Not all design types are equally efficient with respect to heat exchange. The more force-directed and turbulent the flow, the more efficient the heat exchange.

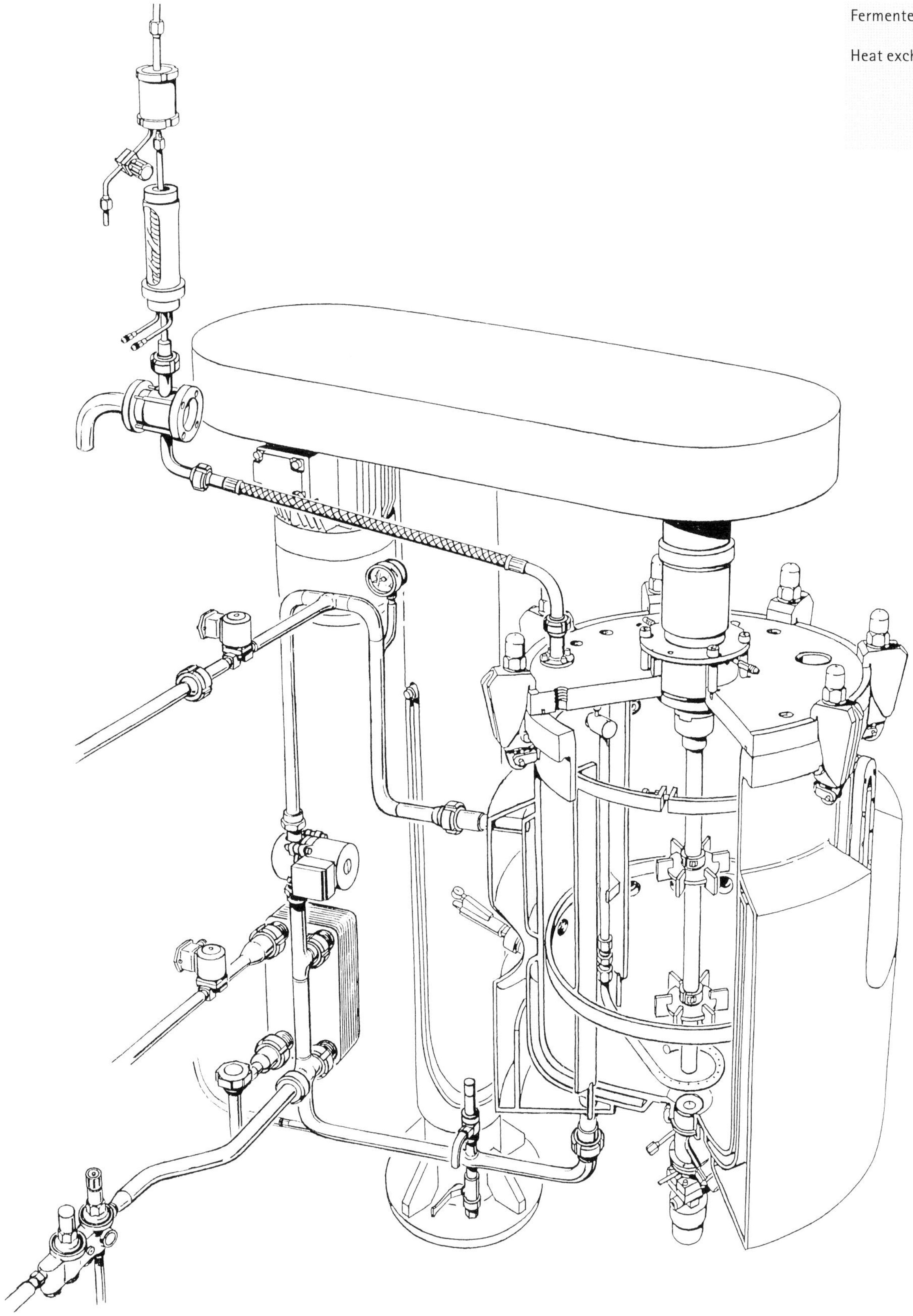

Reactor with a double jacket as heat-exchange surface.

Reactors are often equipped with an insulating jacket. This does not necessarily depend on the presence of a temperature circuit. The insulation is primarily used for temperature control. This also protects operators against the risk of burns. In addition, insulation reduces heat radiation and thus prevents excessive heating of the service rooms. When many reactors are operated in relatively small rooms, this factor should be taken into consideration to see whether additional room cooling is required. Insulation materials may be of various types. Glass wool is most widely used.

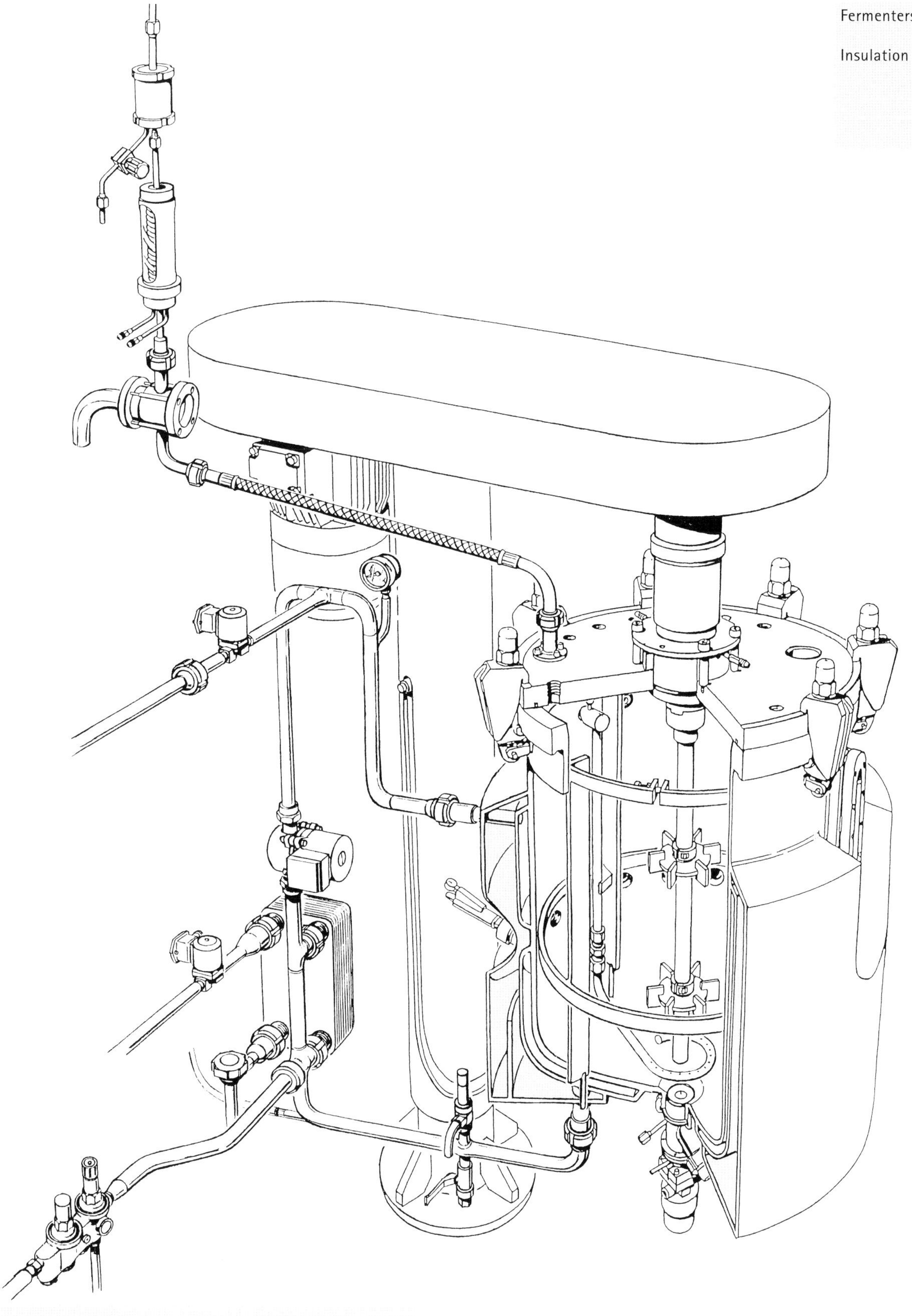

Cutaway view of the insulating jacket of a reactor.

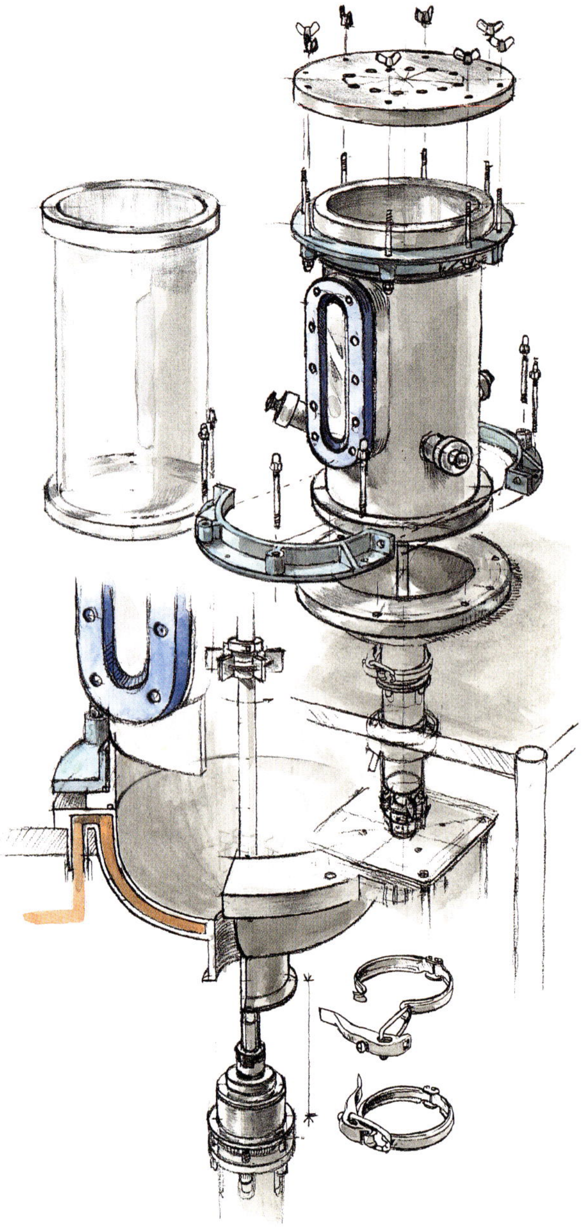

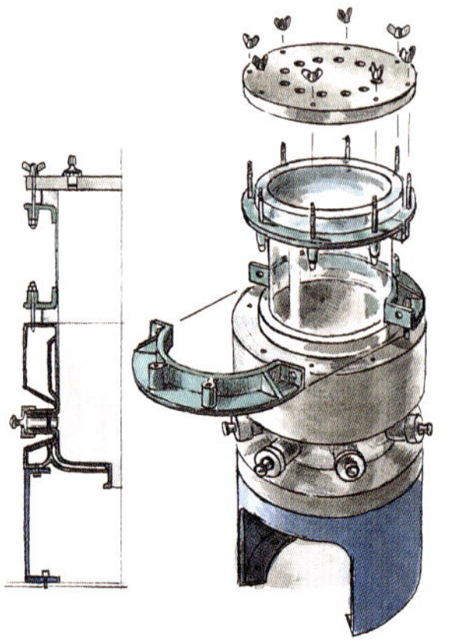

Bioengineering small-lab system in a typical design integrated into a rack (left) or into a vessel base (above)

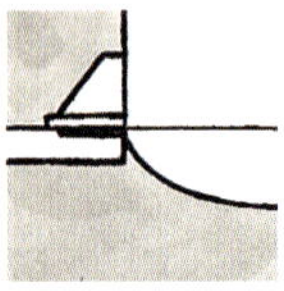

Lug

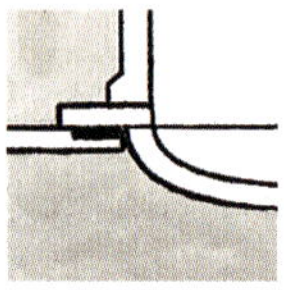

Lug on insulated vessel

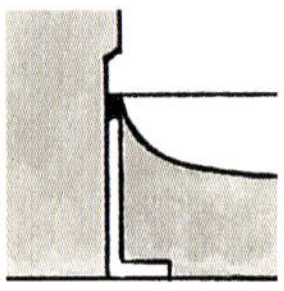

Skirt

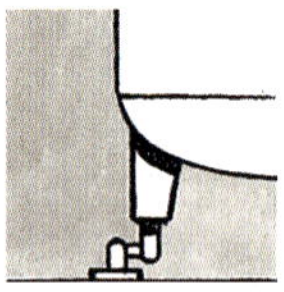

Leg with integrated load cell

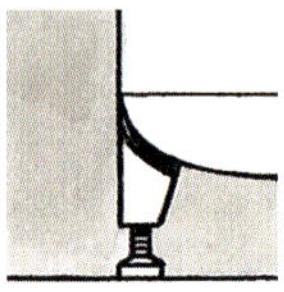

Leg with leveling foot

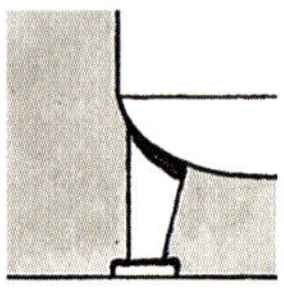

Leg permanently bolted to floor

Every reactor – whether large or small – requires some kind of supporting structure. In most cases, these are legs designed as pipe constructions whose ends are permanently bolted to the floor. Vessels may also be suspended on lugs within the scope of platform frame constructions. Lugs are very space-saving solutions that can be installed in the floor. Another solution instead of legs or lugs are skirts, particularly in case of large vessels. They provide homogeneous load distribution but block access to the lower part of the reactor, which may be a disadvantage. Vessel legs may be equipped with load cells and, if necessary, leveling feet.

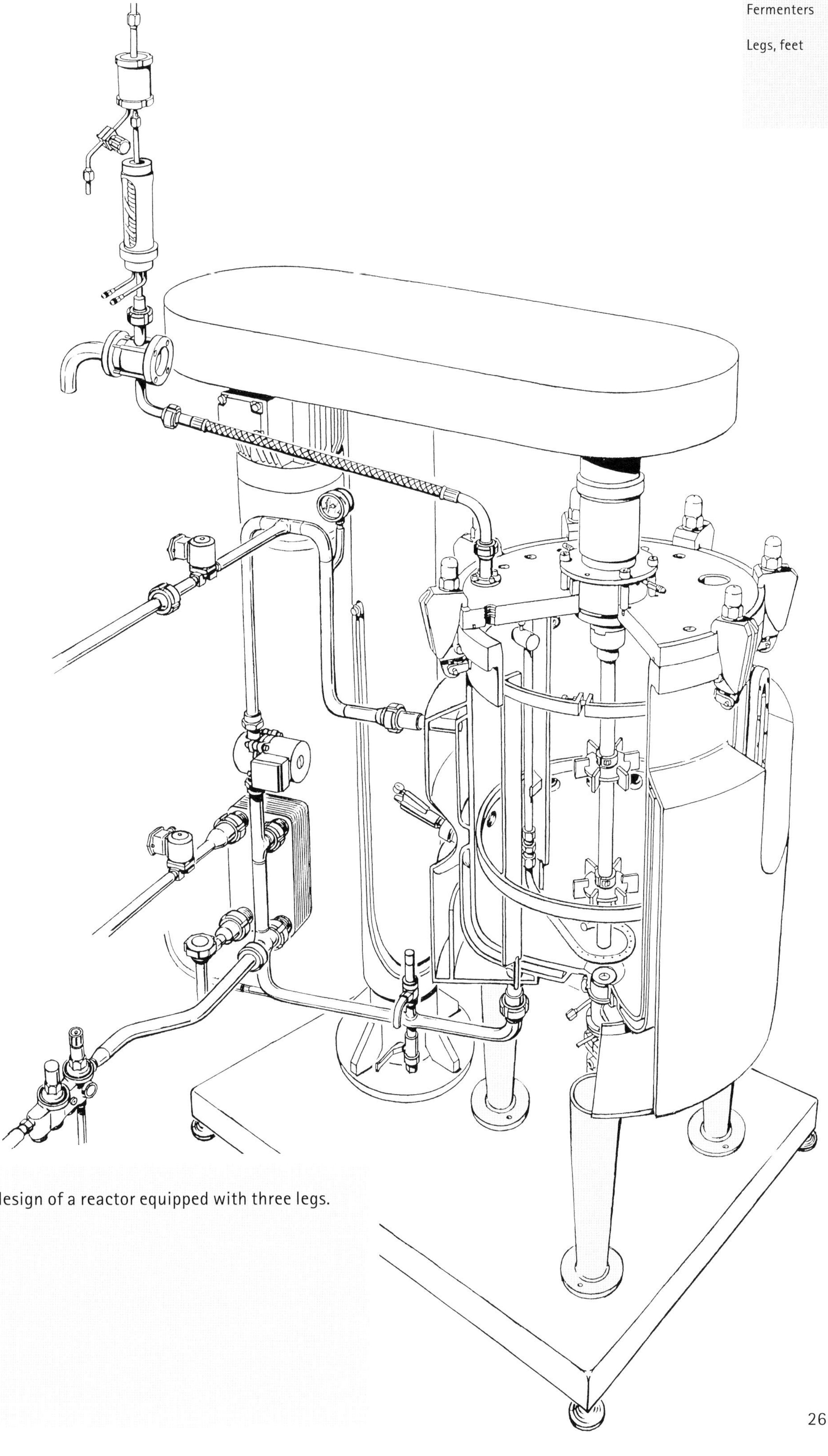

Typical design of a reactor equipped with three legs.

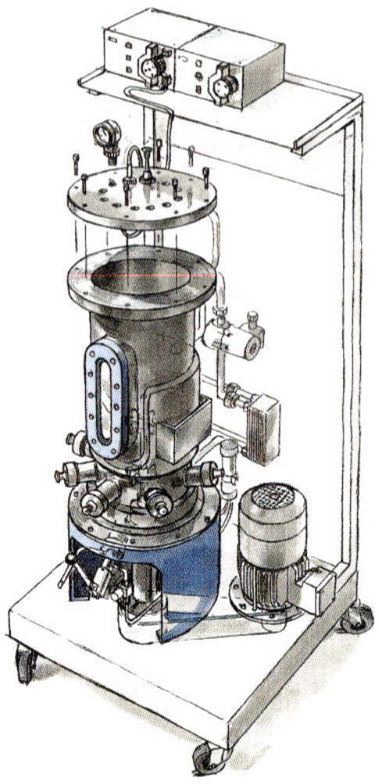

Bioengineering small-lab system in a typical design on a space-saving, mobile platform

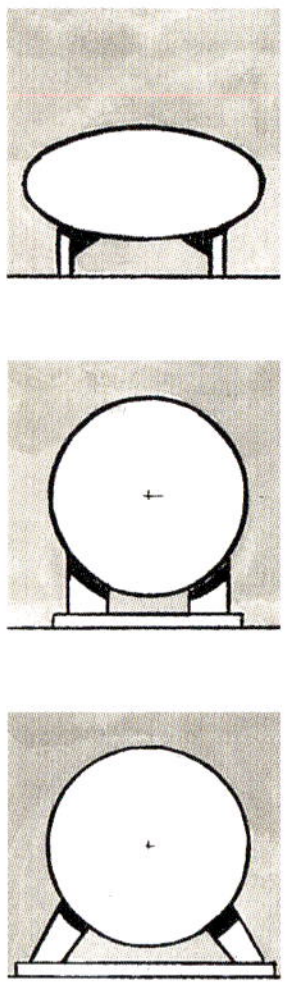

Horizontally oriented vessel on saddles

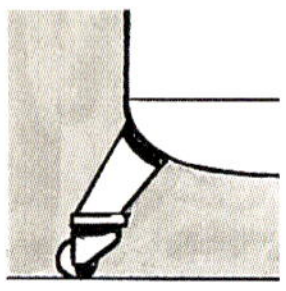

Leg with caster

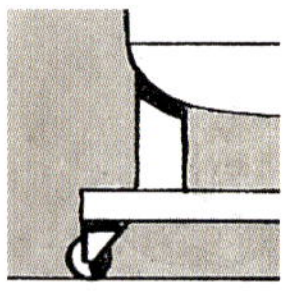

Platform with casters

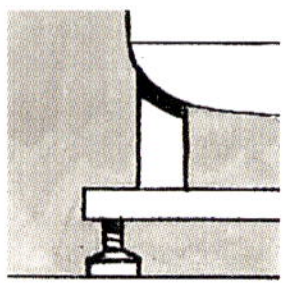

Platform with leveling foot

In addition to conventional leg designs, there are a number of expanded forms of vessel supports such as platforms, wheels, etc. Mobile vessels are equipped with casters. Such designs must ensure that the vessel can be moved without the risk of tipping. If the entire apparatus needs to be mobile – i.e. including the piping – mobile platforms are the solution of choice. Platforms have the general advantage of high stability. Moreover, they can accommodate elements such as load cells, drive components, etc. and protect them against external impact. Horizontally arranged vessels are usually supported on saddles.

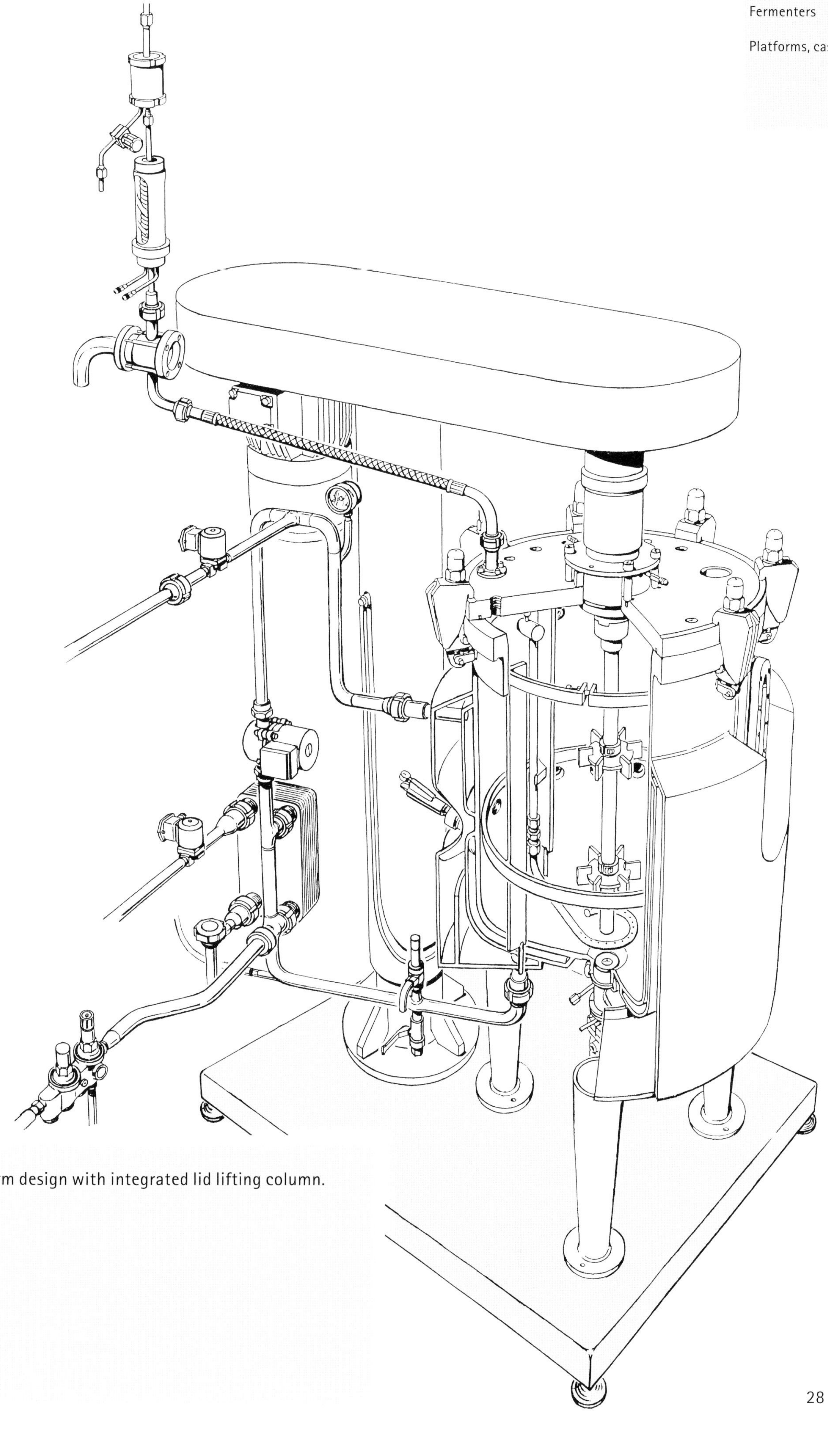

Platform design with integrated lid lifting column.

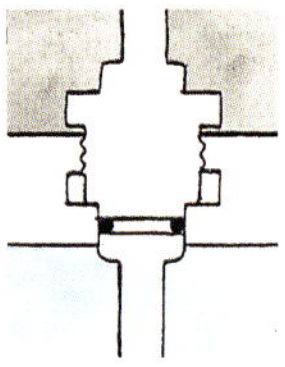

DN19 port

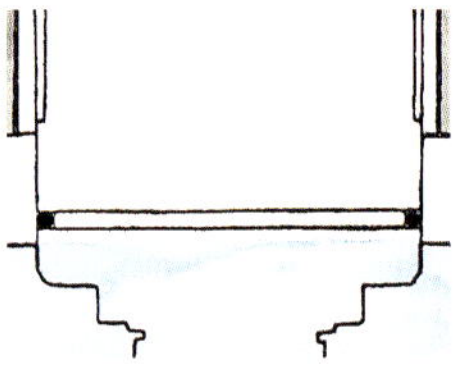

Drive

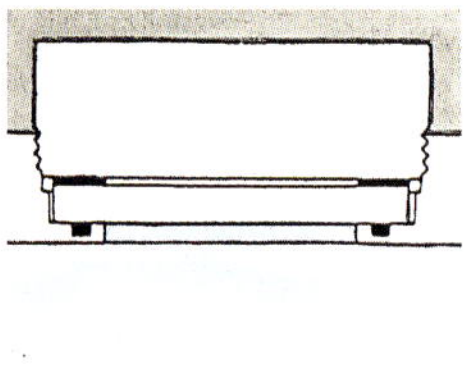

Round sight glass

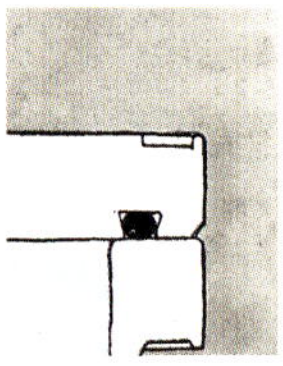

Flat lid

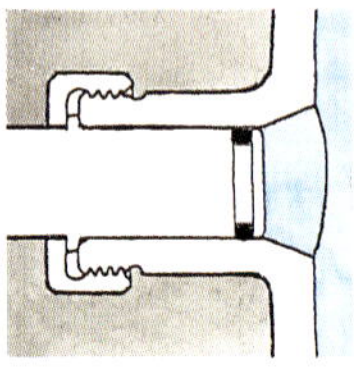

DN25 dummy plug

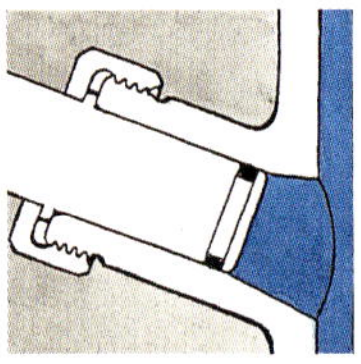

DN25 probe port

Various sealing examples

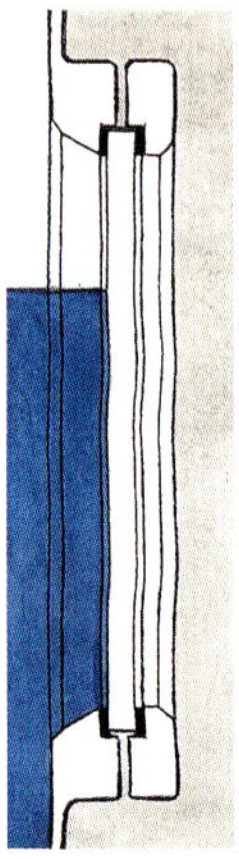

Longitudinal sight glass

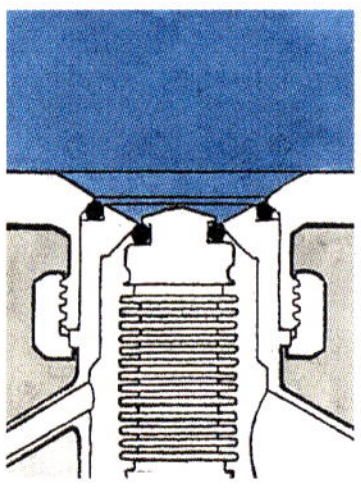

Bottom drain valve

Every detachable connection requires a seal. The greater the number of seals, the greater the risk of contamination through leaks. Sealed connections should not only be perfectly tight but also optimally positioned and easy to service. Non-detachable connections (e.g. welded metering lines) have the disadvantage of being permanently installed, which may cause problems in case of damage (due to mechanical or chemical impact). Detachable, sealed connections can always be retrofitted or replaced when damaged and thus are more flexible. The tightness of sealed connections can be checked with procedures such as pressure-drop tests or leaktightness tests.

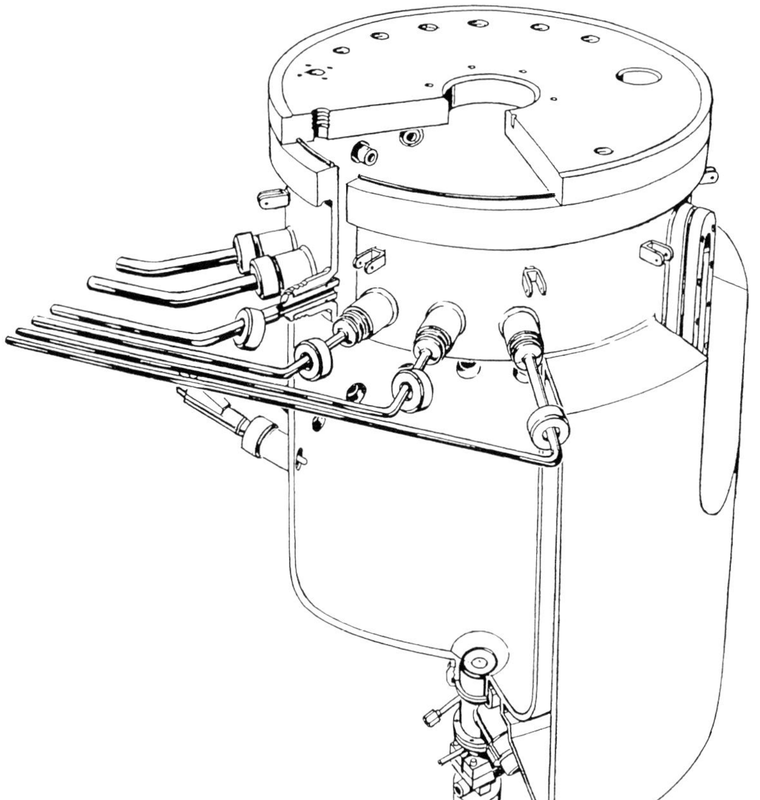

Diagram of various connections on the reactor which are sealed with gaskets.

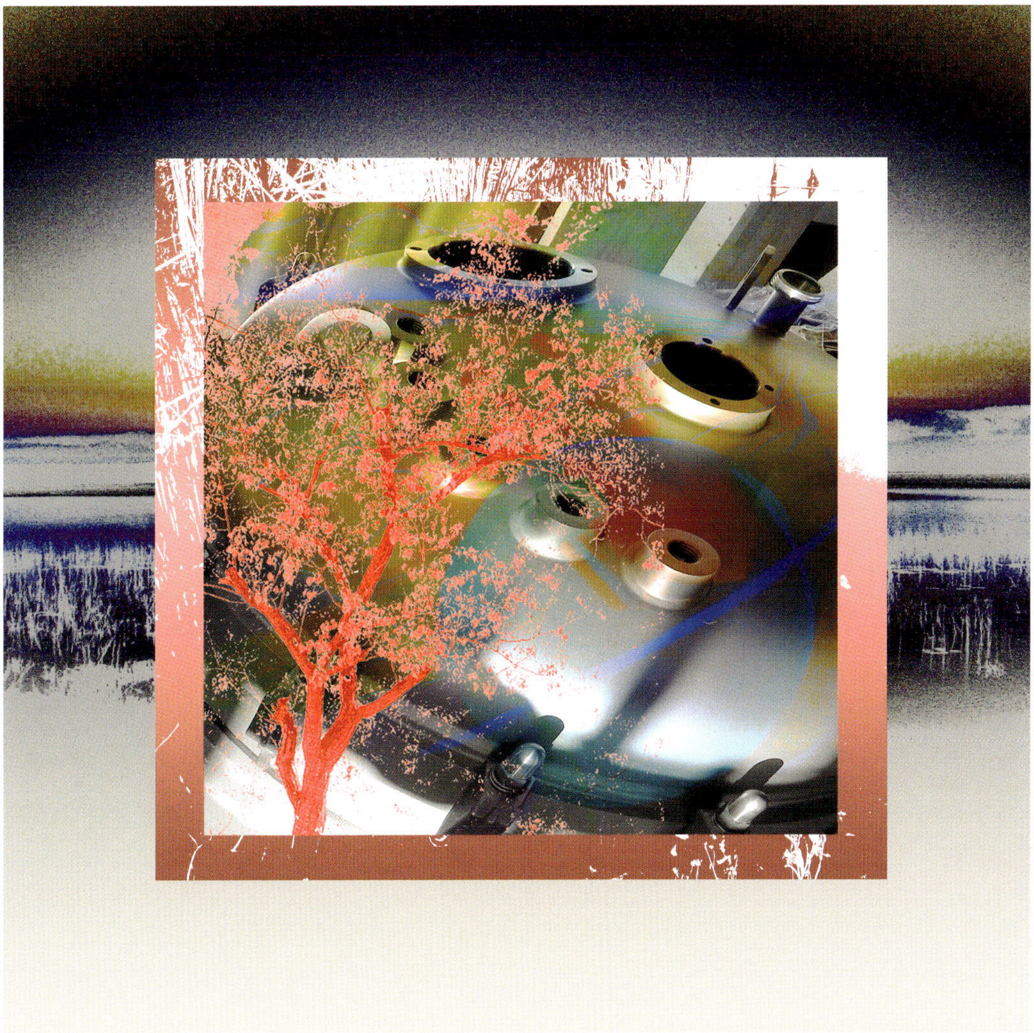

Problem

Solution

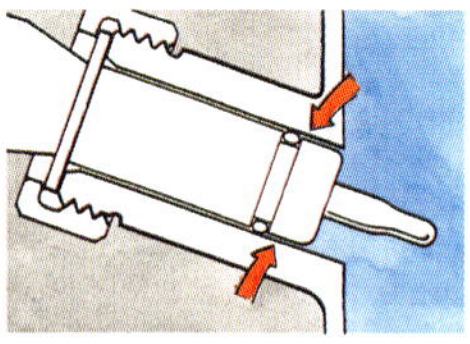

An inclined port entails a gap on the long side.

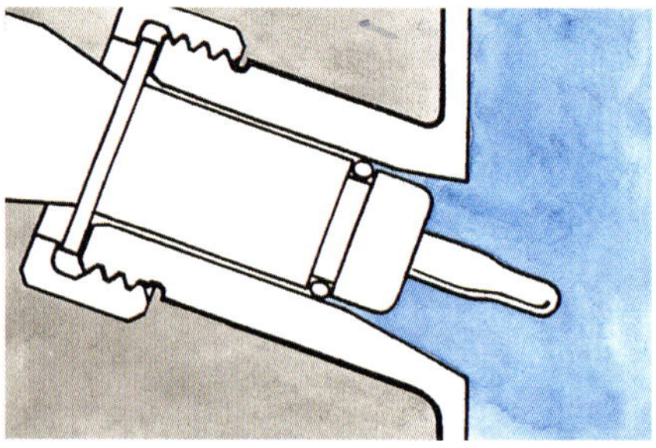

Conical shaping opens the gap.

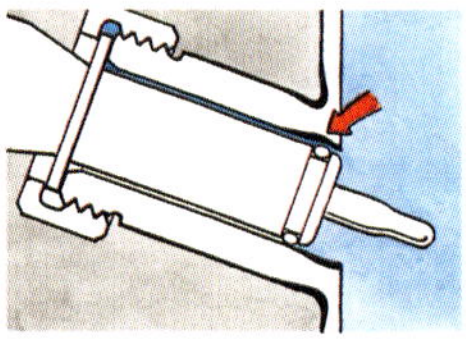

Because heat is applied during welding, the cylindrical bore may be deformed, resulting in a risk of leakage in this area.

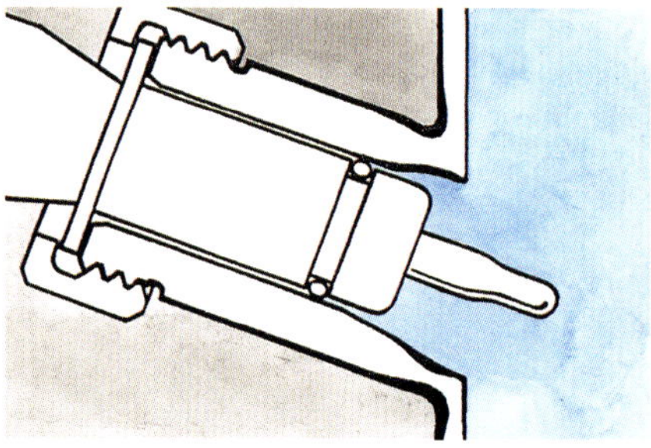

The O-ring is displaced slightly backward, the gap is opened by conical shaping. In the area of the cone, deformations have no consequences.

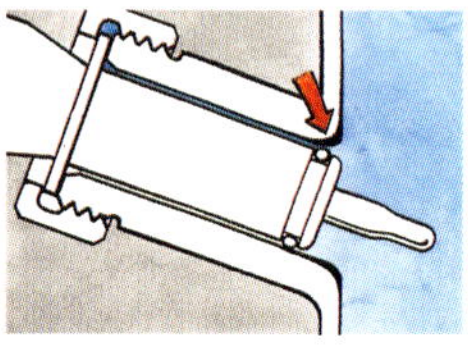

Manual grinding in the sealing area may cause leakage due to excessive removal of material.

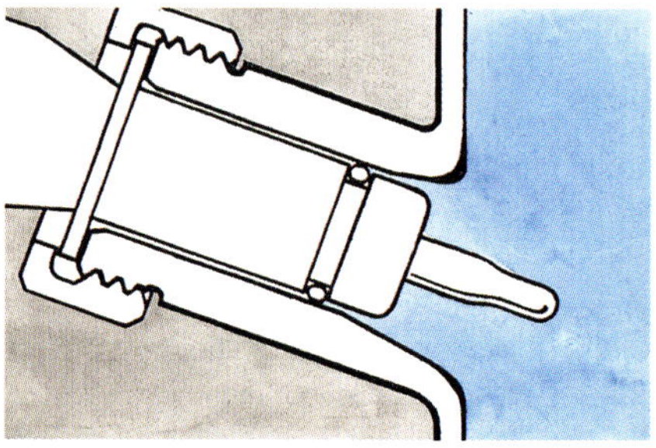

Slightly backsetting the cylindrical area avoids the risk of leakage due to mechanical effects.

Good hygienic design means that gaps and dead spaces are minimized, media can drain, and surfaces can be easily cleaned (CIP). This must always be weighed against other factors such as safety, manufacturing capacity, handling, and costs. The design of a sealed connection should therefore consider not only optimal positioning, but other factors as well. Pressurized vessels are devices that cannot be manufactured in a fully automated process. Many manual steps are necessary. Stainless steel tends to deform under the influence of heat, and this may have a negative effect on the surfaces of sealed connections. For these reasons, the constructive design of a flange with optimal sealing may result in high manufacturing costs. The examples above illustrate some problems that need to be solved in the design of a port.

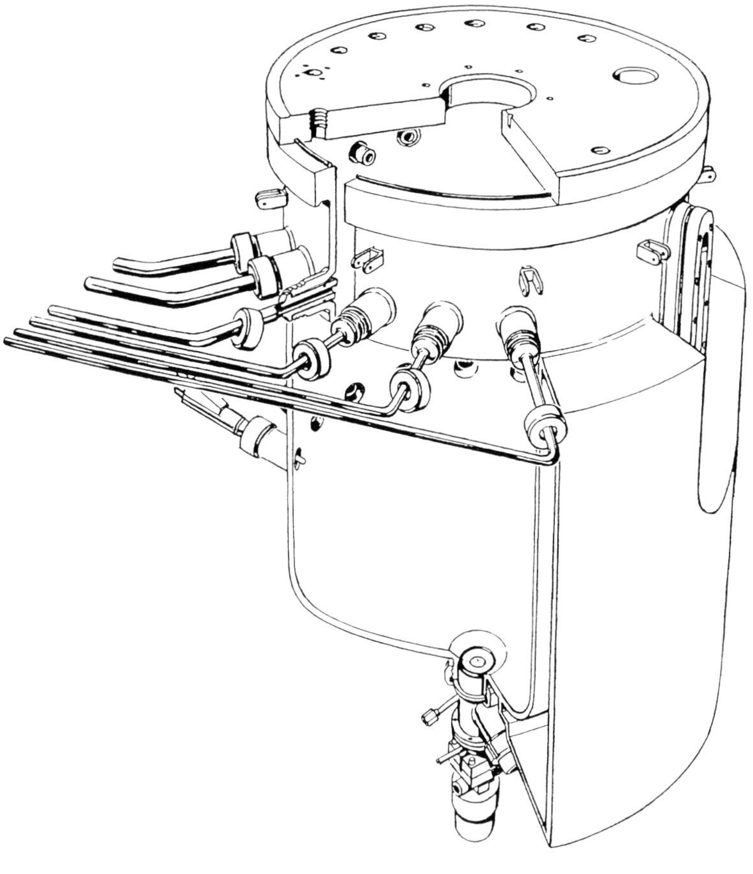

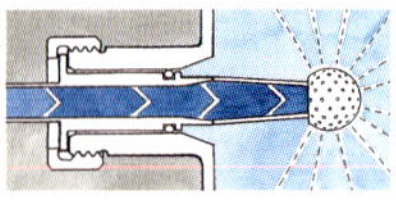

a

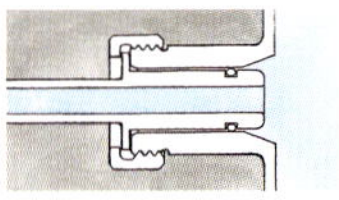

b

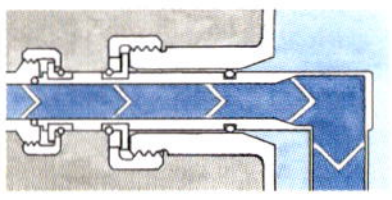

c

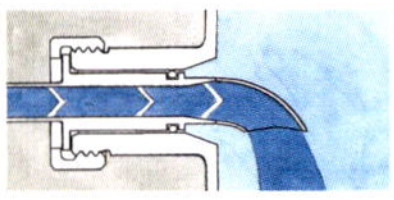

d

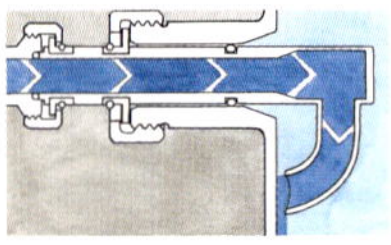

e

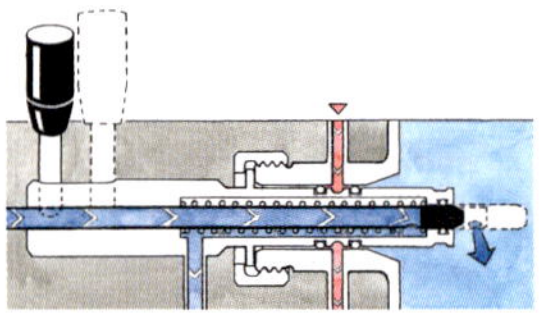

f

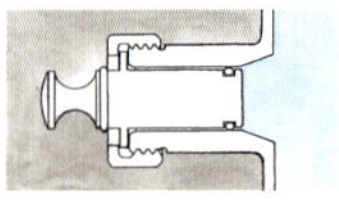

g

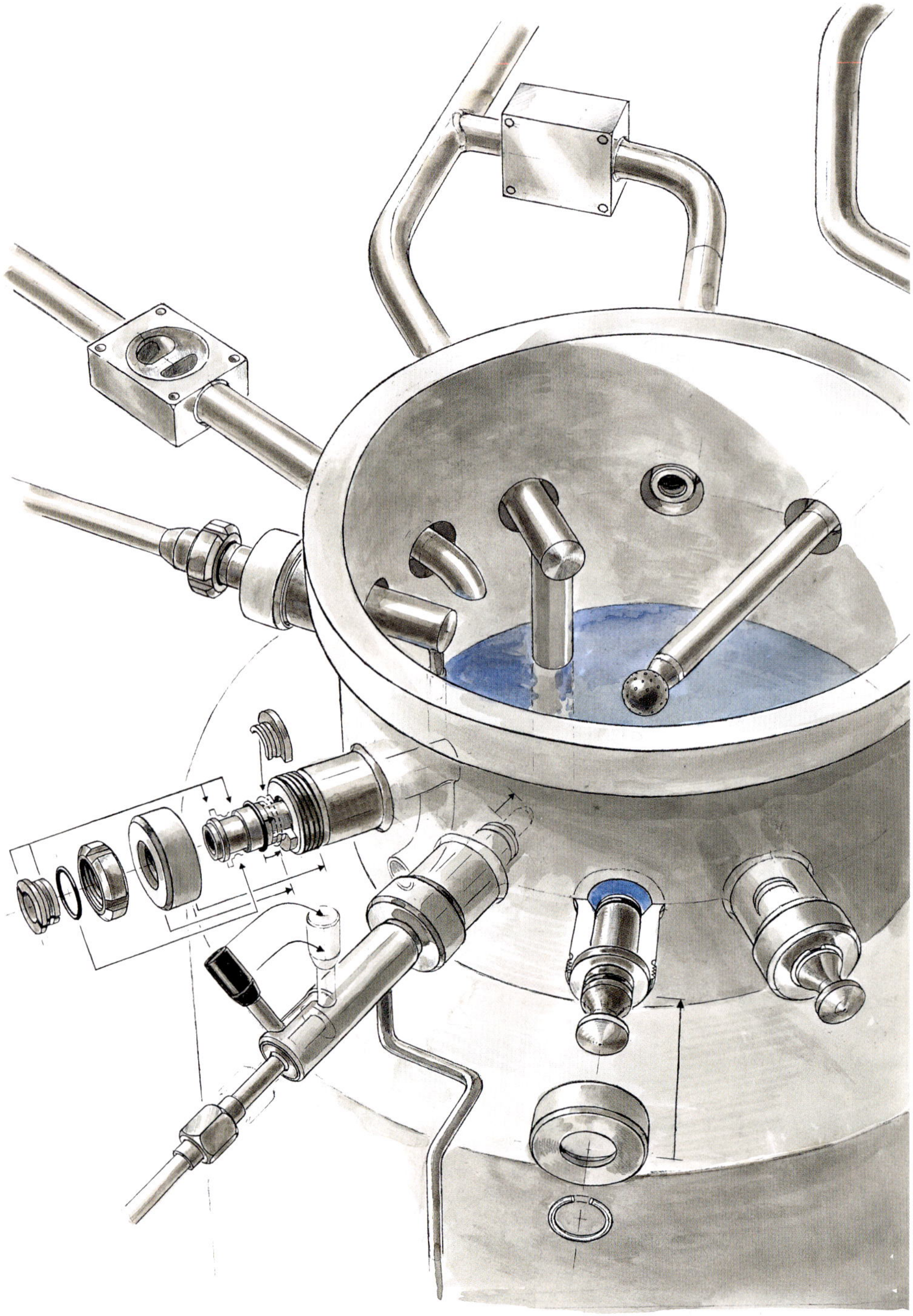

a CIP spherical sparger
b Bypass
c Immersion tube
d Metering line
e Metering line in vessel wall
f Metering valve
g Dummy plug

Metering ports are located on the upper part of the vessel. Lid ports have the disadvantage that all supply and exhaust lines need to be dismantled for opening the lid. Typical applications of metering ports include: metering of alkaline and acid solutions, antifoaming agents, media, or CIP, immersion tubes, supply air, etc. The basis is a standardized vessel port that accommodates all metering ports. Bioengineering metering ports by design contain only the front O-ring near the interior vessel wall but no additional seals or fastening elements in the product-exposed space, thus providing smooth, flush, easily cleaned supply lines to the reactor. Thanks to the high degree of compatibility, retrofitting or upgrading with other ports is easily possible.

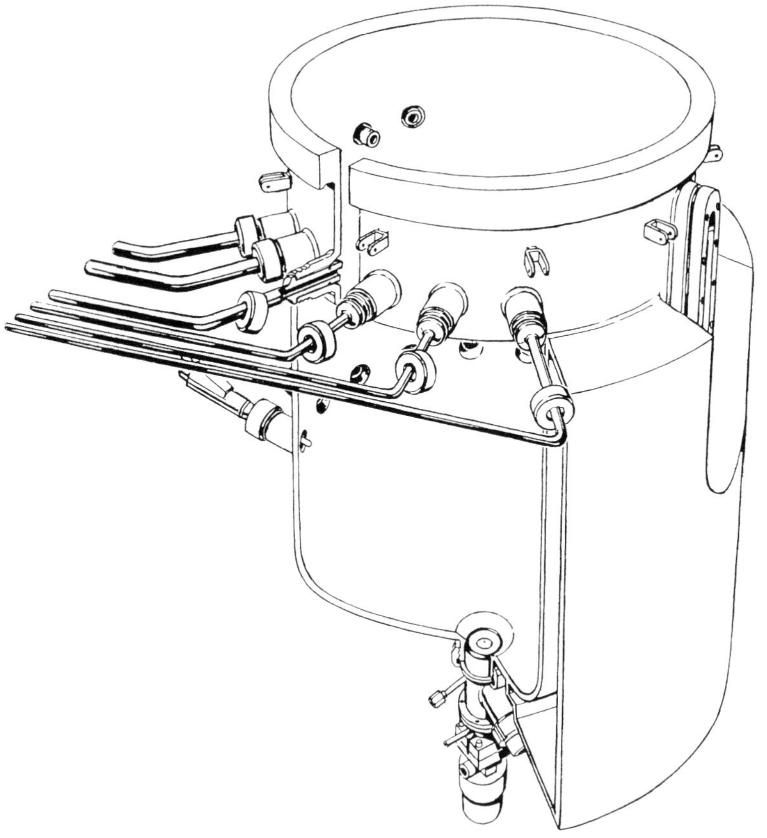

Arrangement of metering ports on a bioreactor.

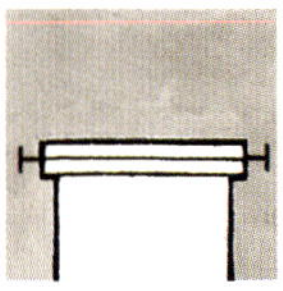

Lid with handles

Lid with lifting lugs

Hinge without spring assist

Hinge with spring package

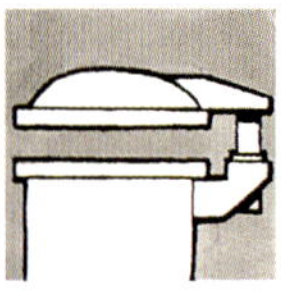

Hydraulic / pneumatic lid lifter

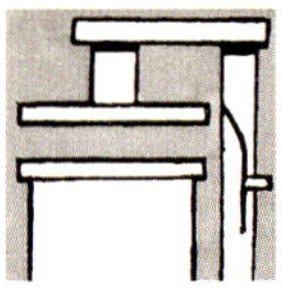

Lid lifting column

Large, heavy lid is lifted by crane on lifting lugs

Medium-sized lid with hinge and spring assist mechanism

Small lid with handles

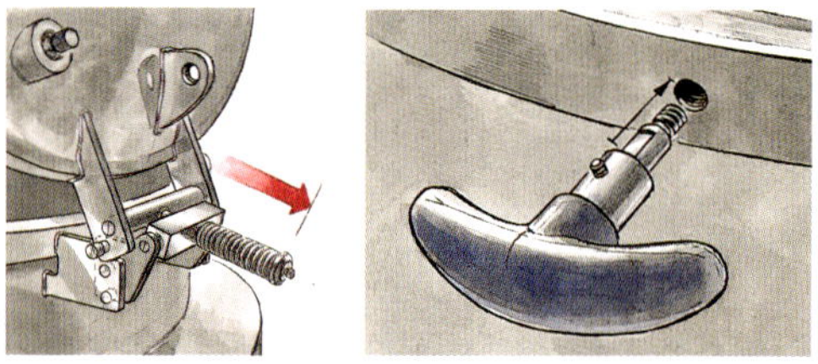

Detail of spring package (left) and handle (right)

Almost all reactors are equipped with a lid, or at least an opening for access to the vessel's interior. Since every lid needs to be opened, this should be a consideration when planning the reactor. Not every location has the equipment (trolley, crane, lifting gear) needed to lift the lid, and retrofitting of such equipment is often impossible. Moreover, space conditions may make it impossible to access the lid with mobile lifting gear. Finally, it is important to account for occupational safety guidelines which may significantly influence the design of the lid.

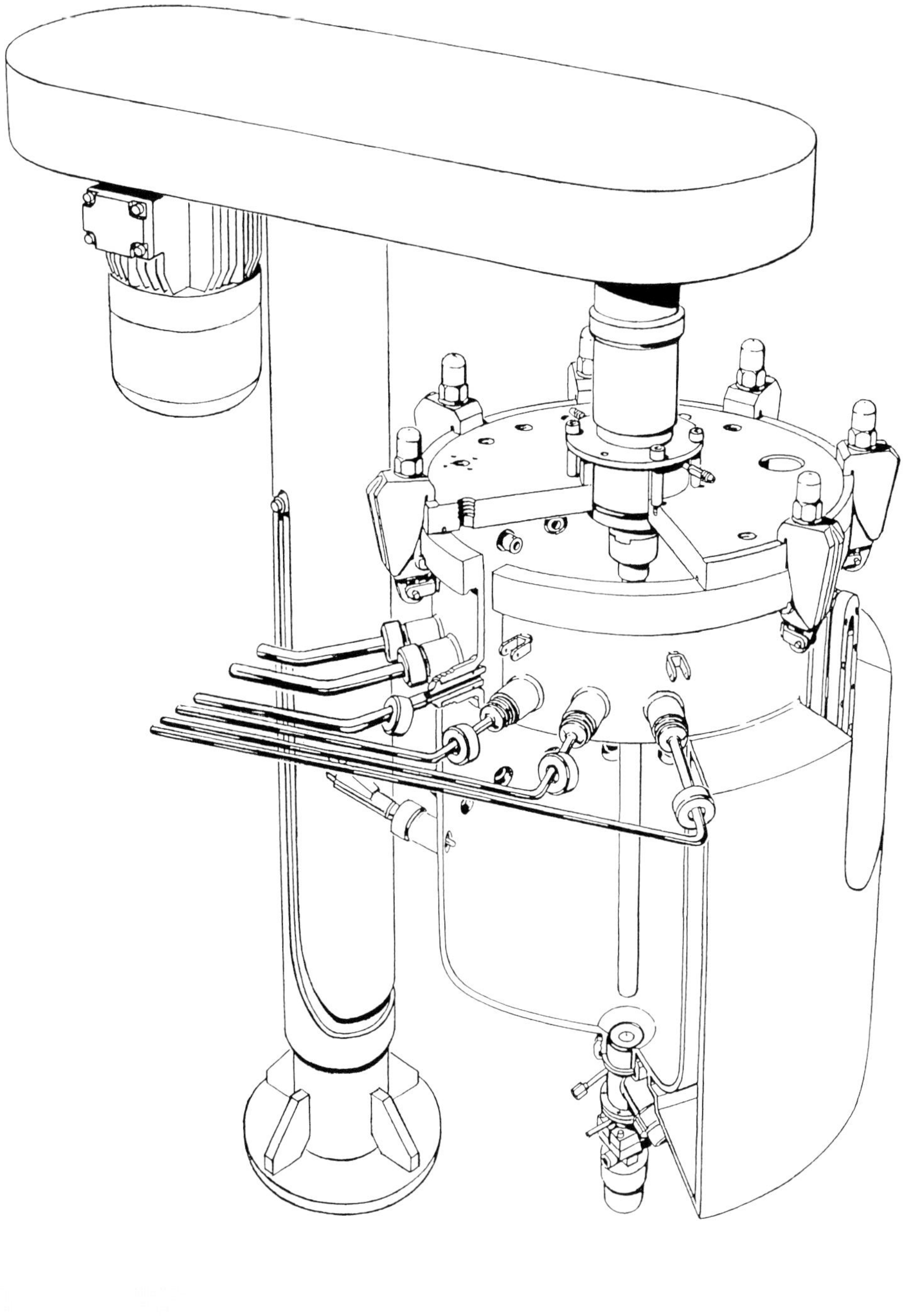

A customized lid lifting column assembly with swivel mechanism.

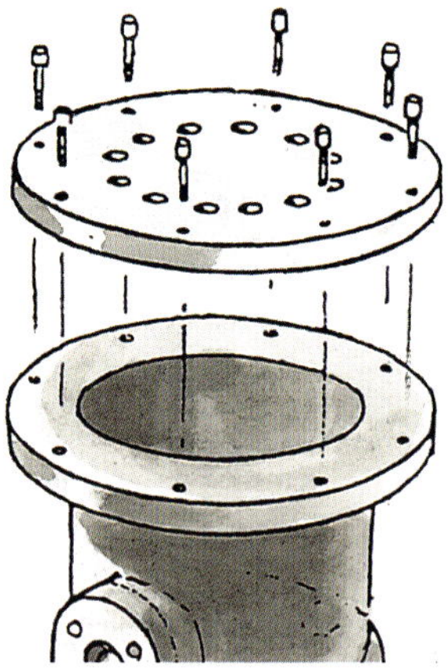

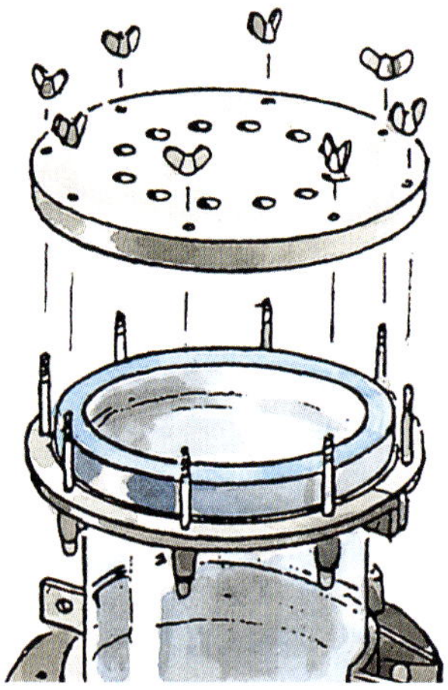

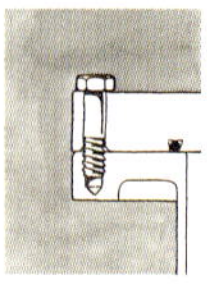

Simple connection of lid and vessel flange with hex-head screws.

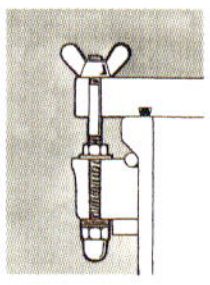

Fastening of lid on captive bolts with wing nuts.

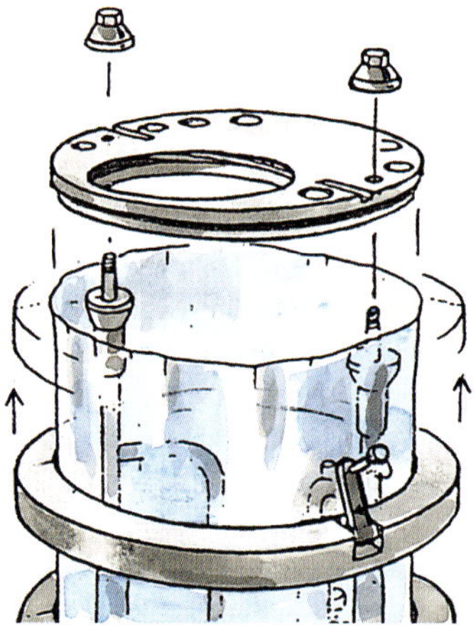

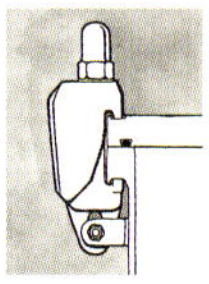

Typical clamp screw solution.

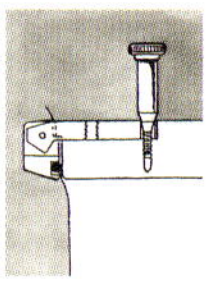

Clamping lock especially developed by Bioengineering closes lid with lid flange and seals foil against lid.

Wherever lids, caps and the like are present, connecting elements are used. From simple screws to segmented clamp screws, a broad range of suitable elements is available. Three basic criteria are always considered: to provide adequate dimensioning (strength), tight sealing (equal distribution, force transmission), and the simplest possible handling. The type and dimensions of a connecting element strongly depend on the design of the lid and the application. In practice, solutions range from lids that are closed manually without any tools to arrangements requiring a visually overwhelming array of segmented clamp screws with precisely defined tightening torques.

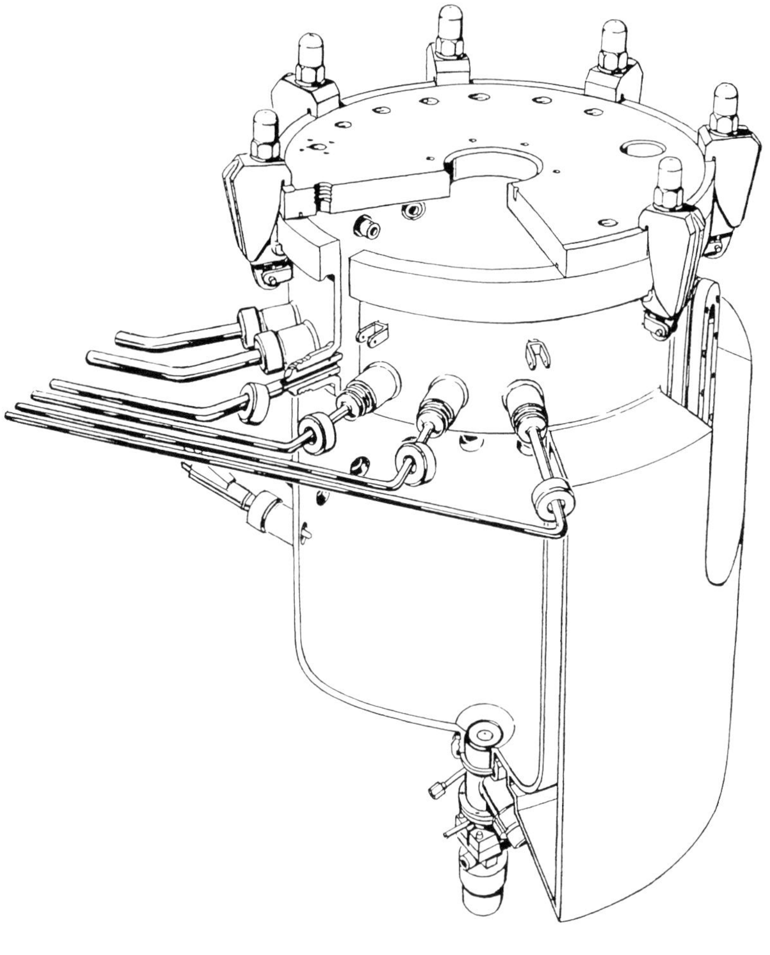

Bioreactor with flat cover and clamp screws as connecting elements.

Vessel without
removable lid

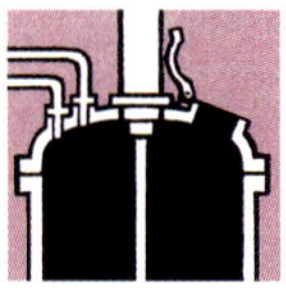

Vessel with
removable lid
and manhole

For vessels exceeding a certain size, removable lids are often impractical because they would be too large and thus difficult to operate. Moreover, lid flanges require a great amount of manufacturing work and this results in high costs. To provide the necessary access to such vessels, they are equipped with access openings called manholes. The manhole should always be designed large enough to enable convenient entry into the vessel and to facilitate the introduction of internal components (drive shaft, stirrers, etc.) into the vessel.

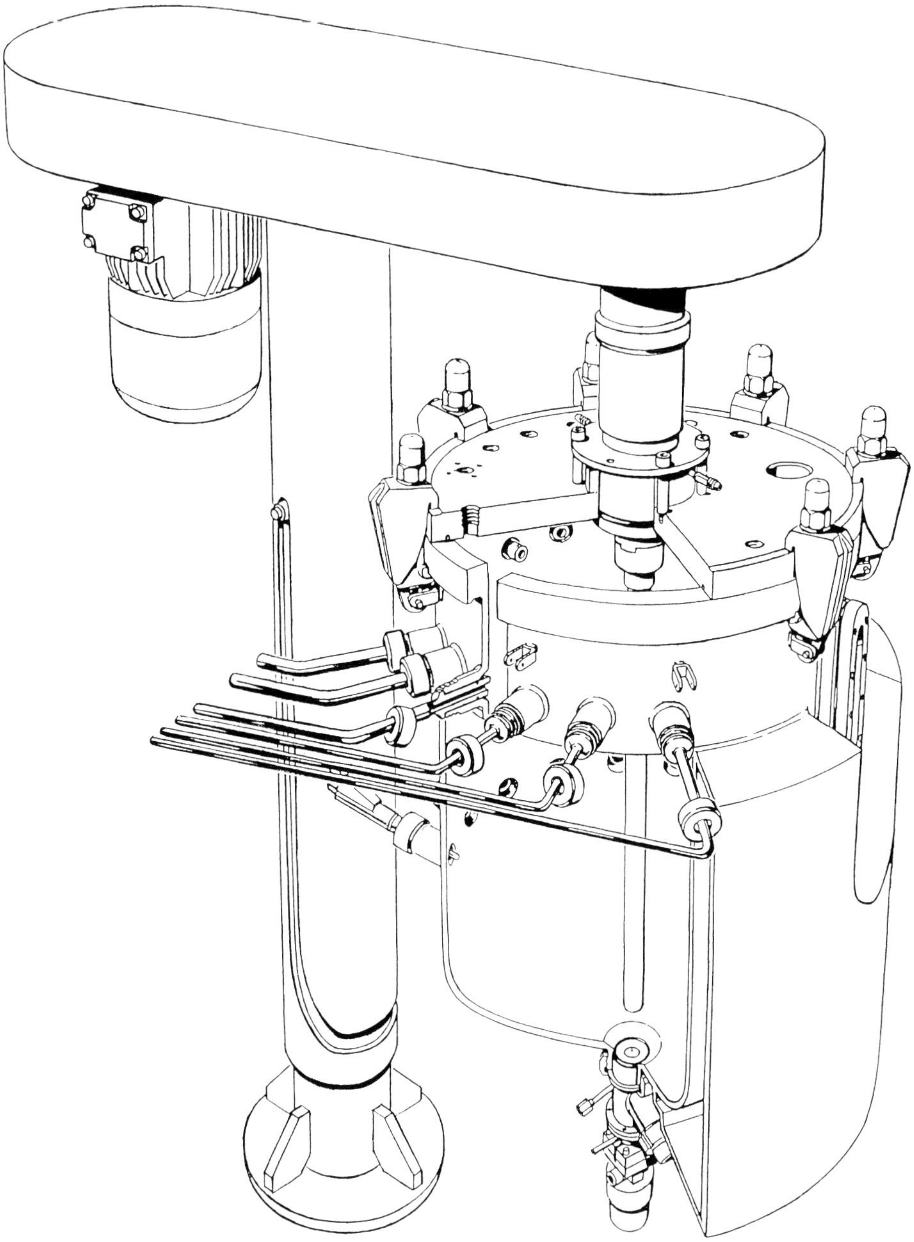

Small vessels are not equipped with manholes.

Pressure sensor and pressure gauge.

Safety valve.

Various lid ports for immersion tubes, exhaust air, etc.

Sight glass and illumination device. Combinable with various sight glass types.

Lid ports are indispensable because every reactor is equipped with fittings and components for the performance of various tasks in the headroom area. For this purpose, it is important that sufficient flexibility be provided. The Bioengineering DN19 port can accommodate numerous components in a very small space, which makes adaptations or retrofitting of equipment simple and straightforward. Moreover, additional lid ports are designed as modular units that can be used in a variety of applications. Lid attachments should be easy to disassemble, take up little space, and be flush with the inside surface.

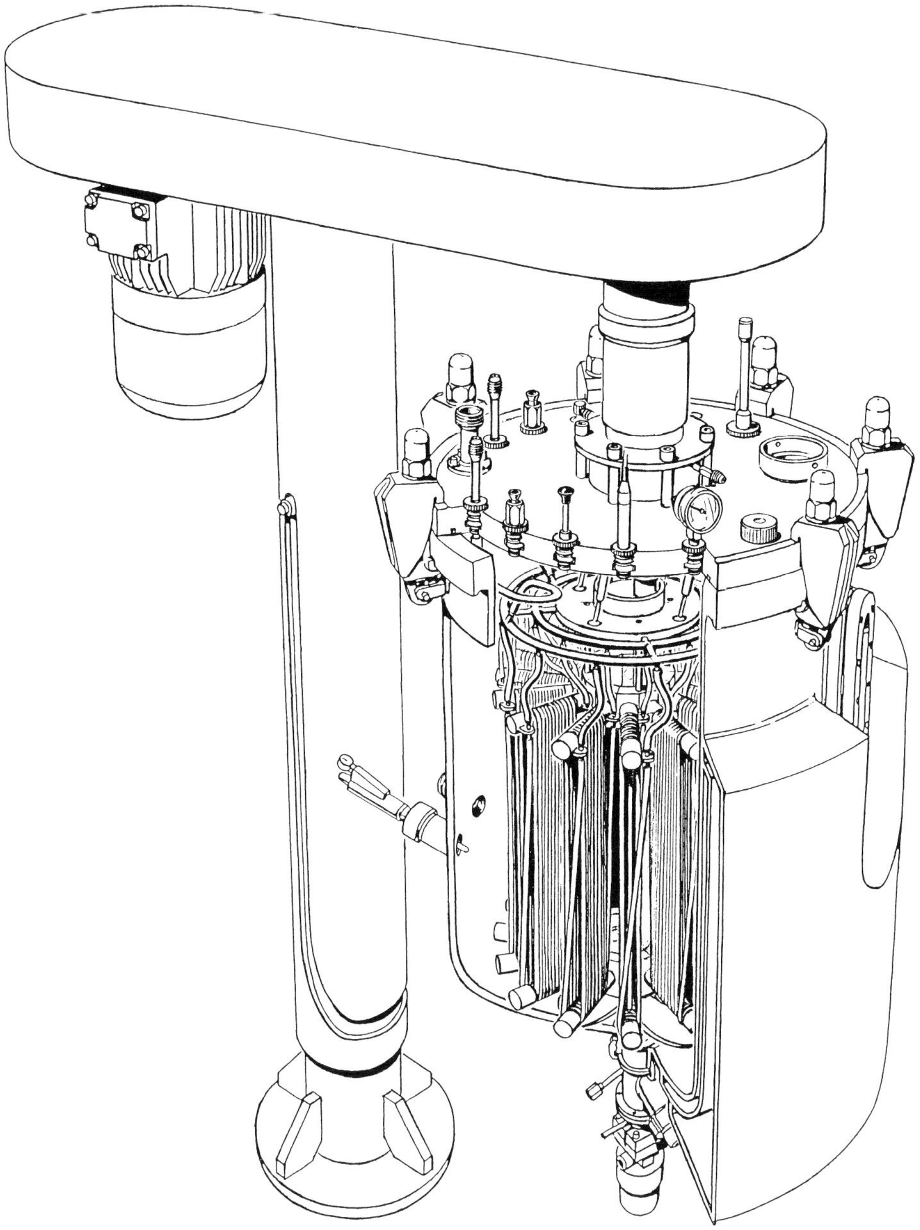

Example of a lid with various built-in components. Space-saving, easily accessible, and designed for assembly and disassembly without tools.

Reciprocating mixing movement

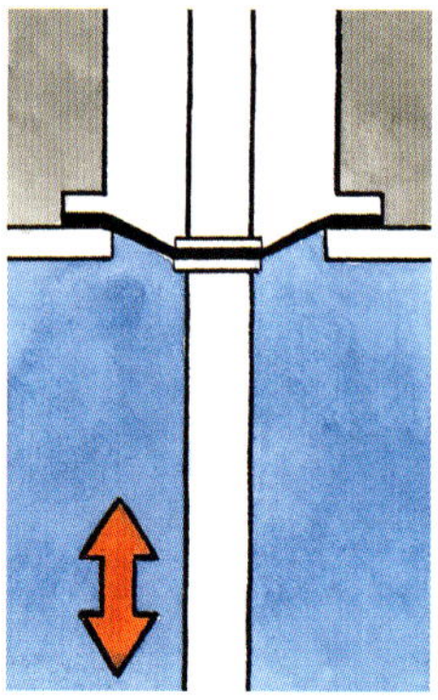

Diaphragm seal

Loop-type mixing motion (nutation)

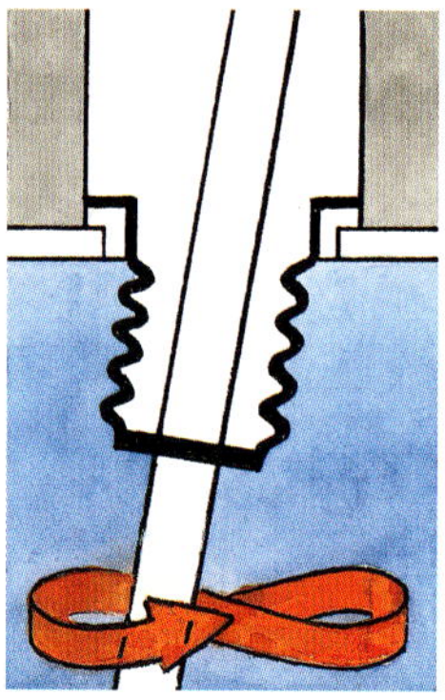

Bellows-type seal

Rotary mixing movement (magnetic stirrer)

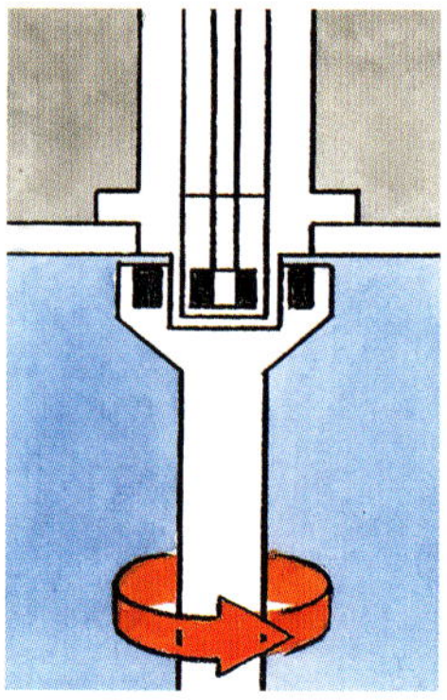

Hermetic seal (encapsulation)

Rotary mixing movement

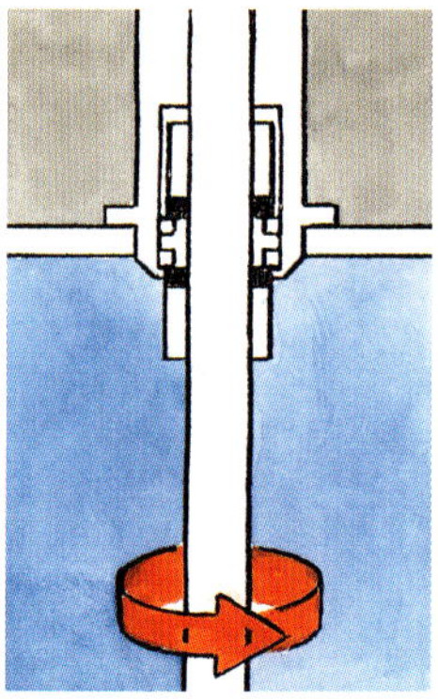

Double mechanical seal

Whenever a medium is agitated by a stirrer, a movement must invariably occur which has to be transmitted to the sterile space of a reactor. Often, it is necessary to seal off the reactor in order to avoid contamination of the medium. Mixing movements are usually of the rotary kind. In practice, the rotating elements are sealed with mechanical seals. Magnetic drives are preferred for sensitive media. They contain no dynamic seals, and the risk of contamination is substantially smaller. Special movement types, such as , reciprocating, oscillating or loop-type movements, are used less frequently. In these cases, the sealing elements include diaphragms, bellows, etc. Sealing elements must withstand physical forces such as pressure, temperature, and speed of movement. They must be resistant to chemicals and be optimally designed to meet all hygienic requirements.

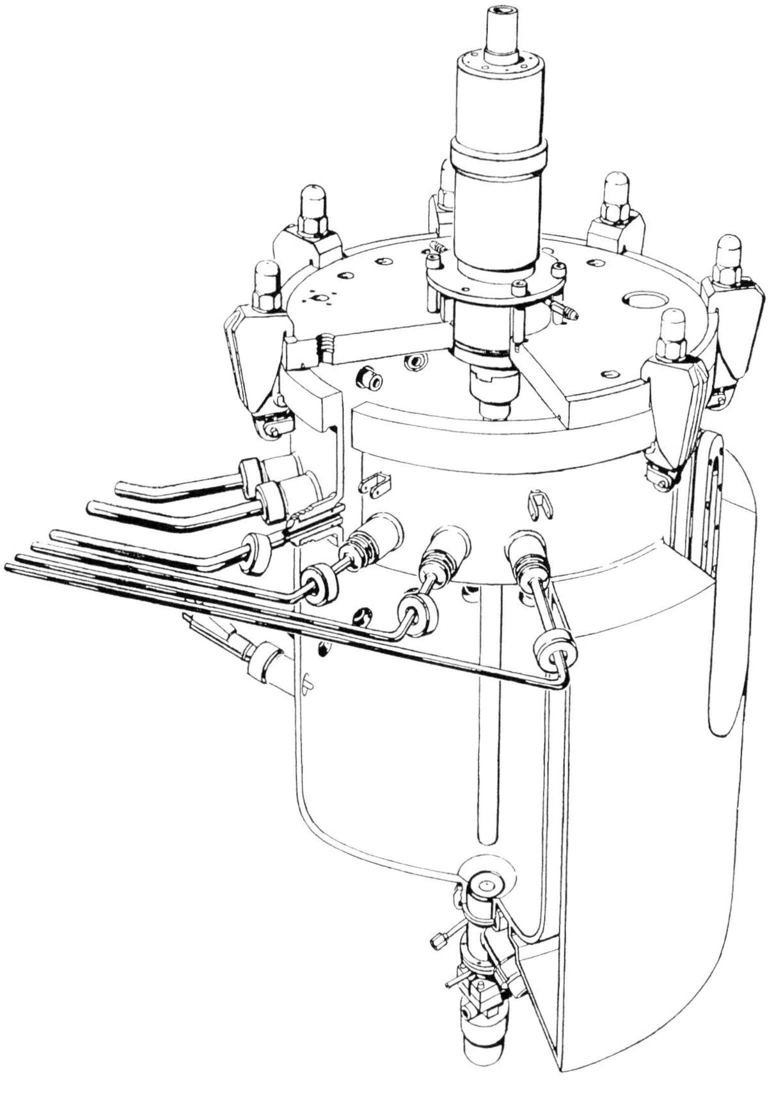

Shaft bearing, doubly sealed with mechanical seals,
to generate a rotary motion in the reactor.

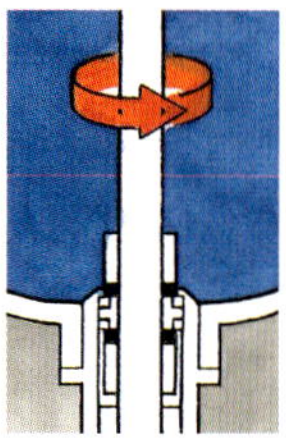

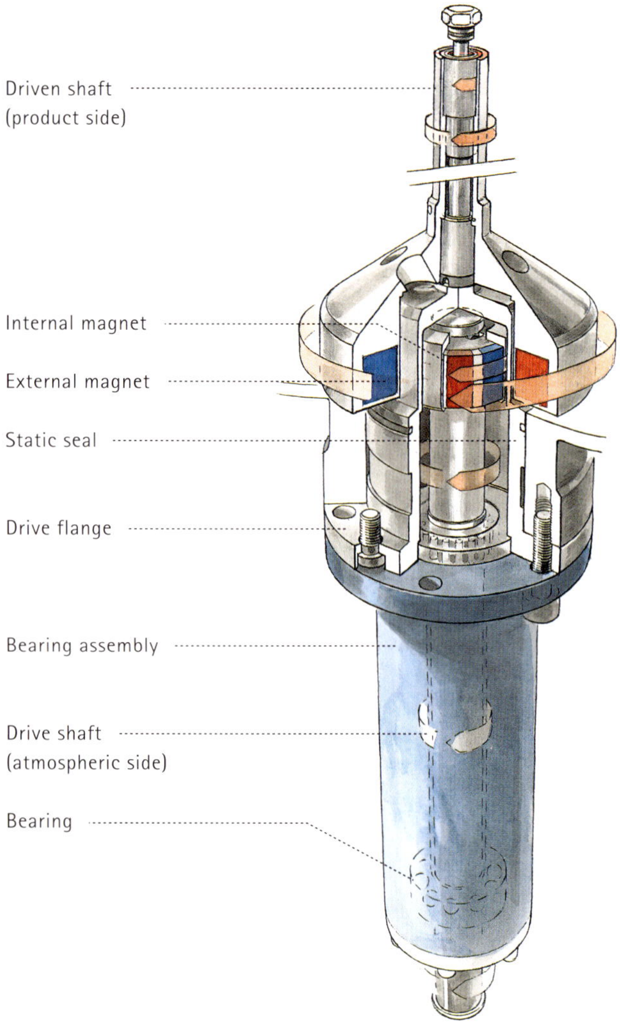

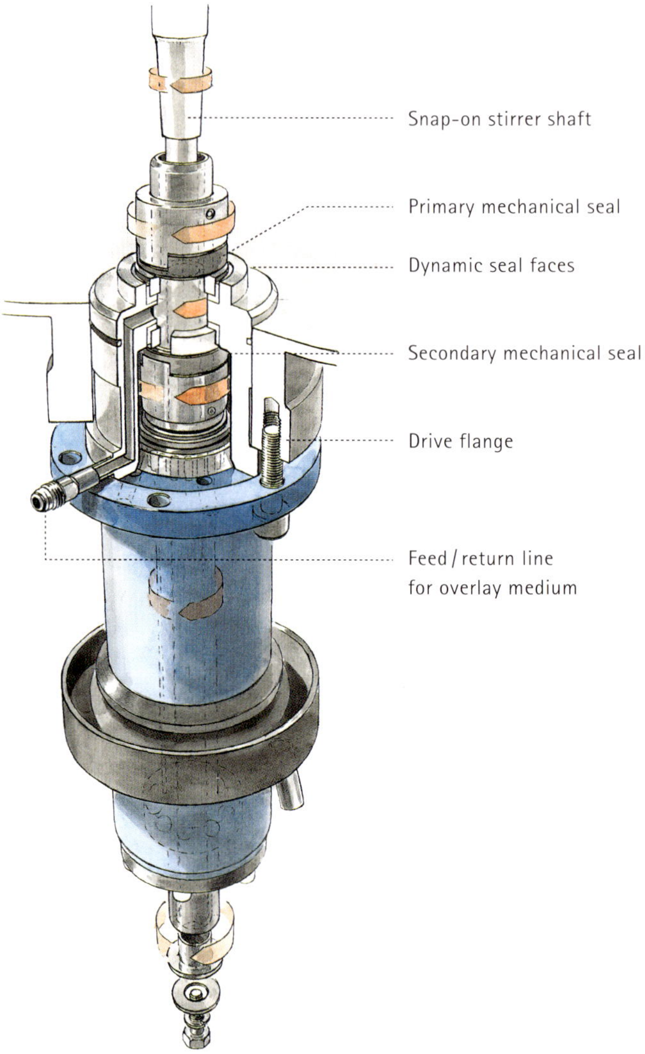

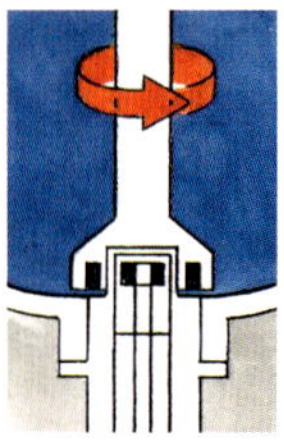
Principle of a magnetic stirrer

Every stirrer function requires an appropriate stirrer system which basically comprises the following components: drive (motor), shaft bearing, stirrer shaft, and impeller. The core element of every drive is the shaft bearing. The stirrer movement is not necessarily rotary, but rotary motion is chosen for most processes. There are several ways in which a shaft bearing may seal off the stirrer shaft from external air. Two principles have become standard: the magnetic stirrer, which seals statically, and the stirrer with mechanical seals, which seals dynamically.

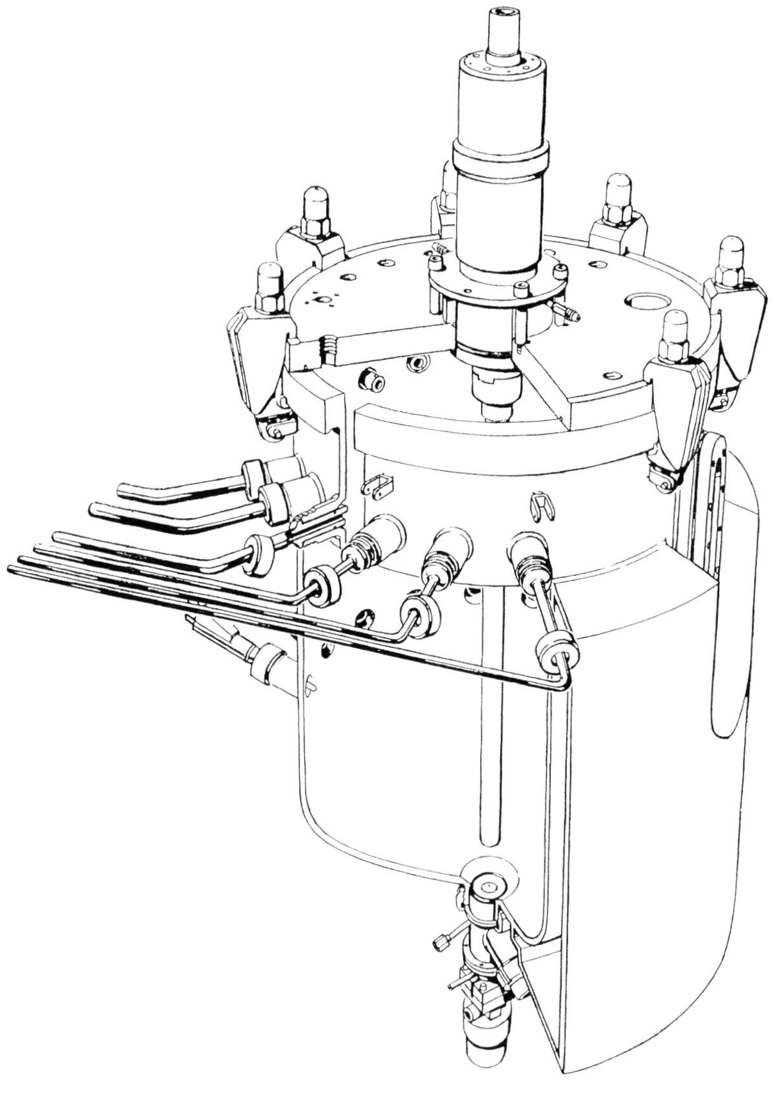

Magnetic stirrer: no fluid barrier required; low risk of contamination; low to medium torques; low wear. Shaft bearing with mechanical seal: no bearing in the medium; flexible as regards shaft design; high torques; versatile.

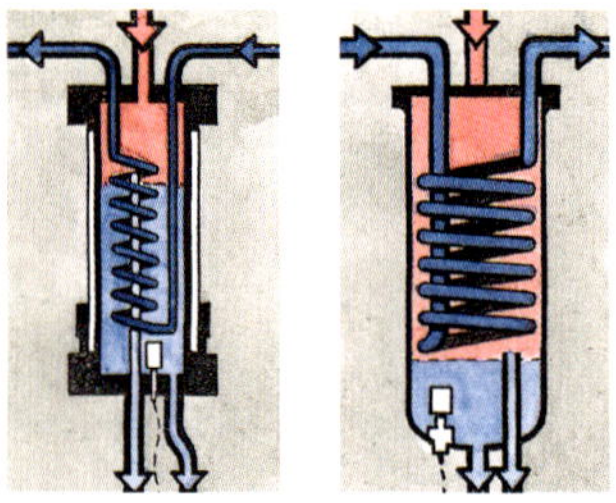

Pressurized condensate vessel.
Left: sight glass design, right: steel vessel. Both with float switch.

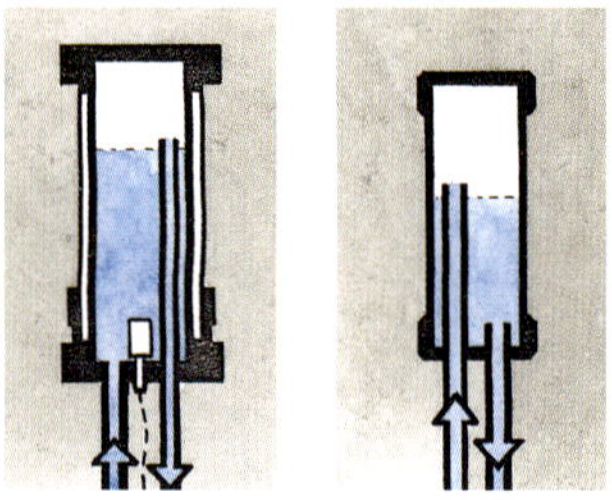

Non-pressurized balancing vessel, with and without float switch.
Manual refilling.

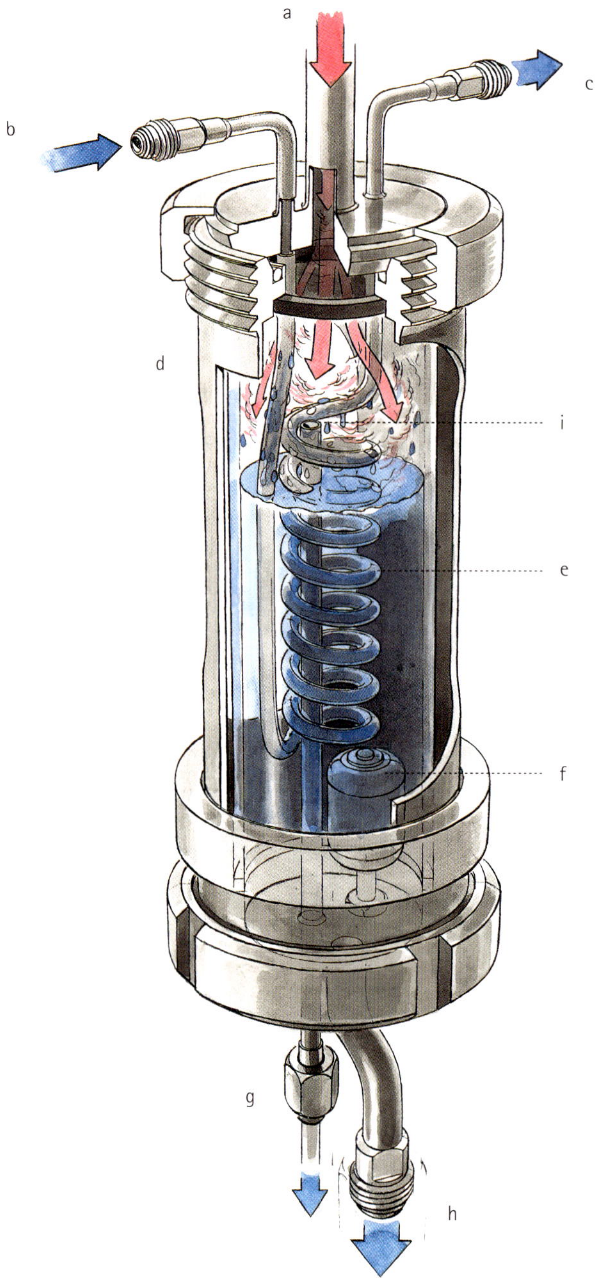

a Steam feed / pressurization with gas
b Coolant inlet
c Coolant outlet
d Condensate vessel
e Cooling coil
f Float switch
g Steam trap
h Condensate
I Overflow

Depending on the task, various systems can be used to balance mechanical seals. In practice, a distinction is made between pressurizing, syphon, or circulation systems. The main functions of a mechanical seal buffering or barriering system include providing a lubricating film between the sliding faces, maintaining a constant sealing liquid pressure, and cooling the mechanical seal. One option is the use of condensate as a sealing medium. The condensate forms in a condensate vessel and is connected with the barrier fluid chamber of the mechanical seal. To ensure continuous balancing with a barrier fluid, there are systems with purely visual level monitoring (sight glass) and systems that are electronically monitored.

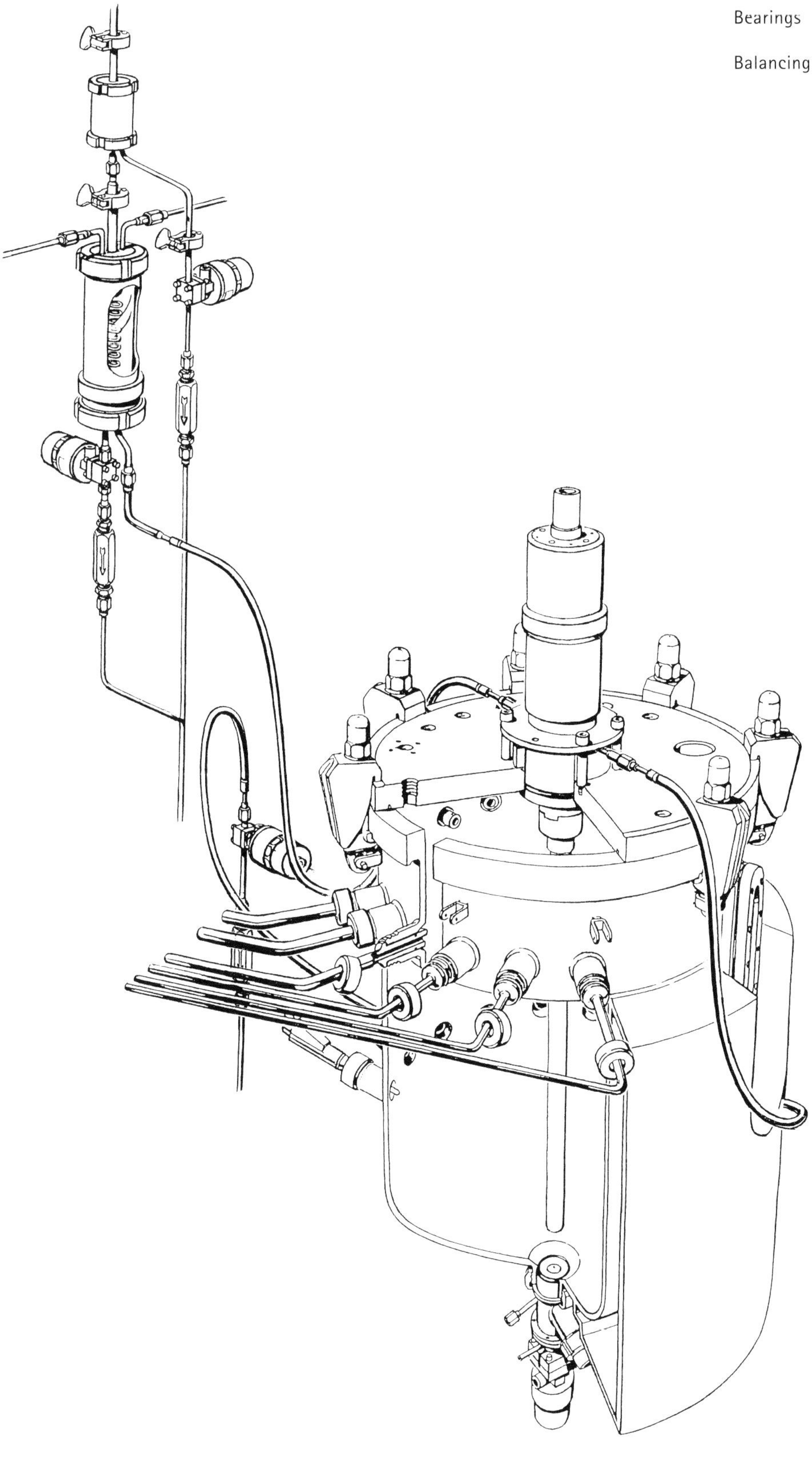

Typical application of a condensate reservoir on a bioreactor.

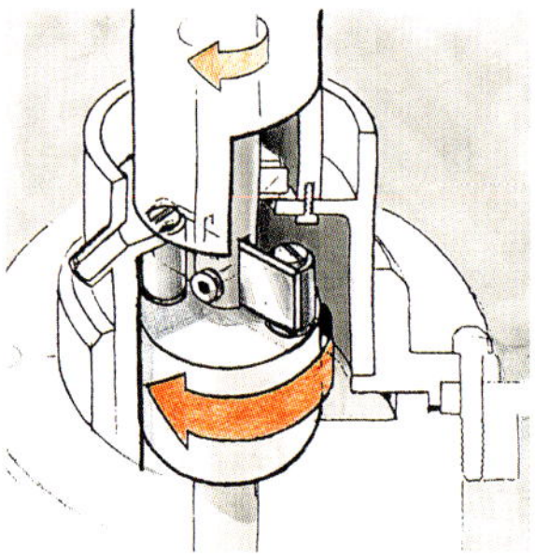

Example of a Bioengineering bearing with hydrostatic coupling as a top-mounted drive; decoupling is simple and requires no tools.

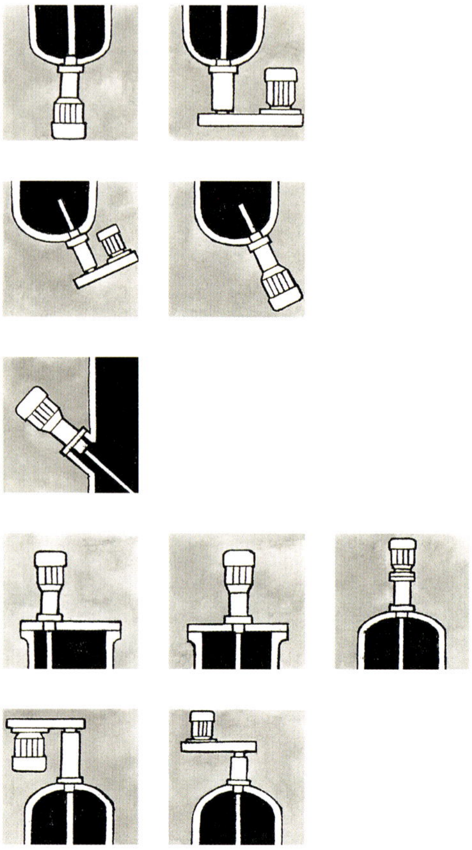

Bioengineering's proven concept is shown in this drawing of a type NFL lab reactor. After disassembly of the lid, the shaft can be separated from the underside drive and removed without tools.

There are various ways to position drives in a reactor. Drives may be mounted at the bottom, at the top, or at the side. The position is often determined by given circumstances. These may include: available space below and above the vessel, stirring method, access for operation and maintenance procedures, geometry and design of the reactor, equipment installed inside the reactor, etc. Especially in larger vessels, positioning is critical in order to facilitate the simple assembly and disassembly not only of the drive but also of the stirrer shaft. Years of experience have shown that the snap-on shaft design developed by Bioengineering is an excellent solution with regard to simplicity of maintenance and operation.

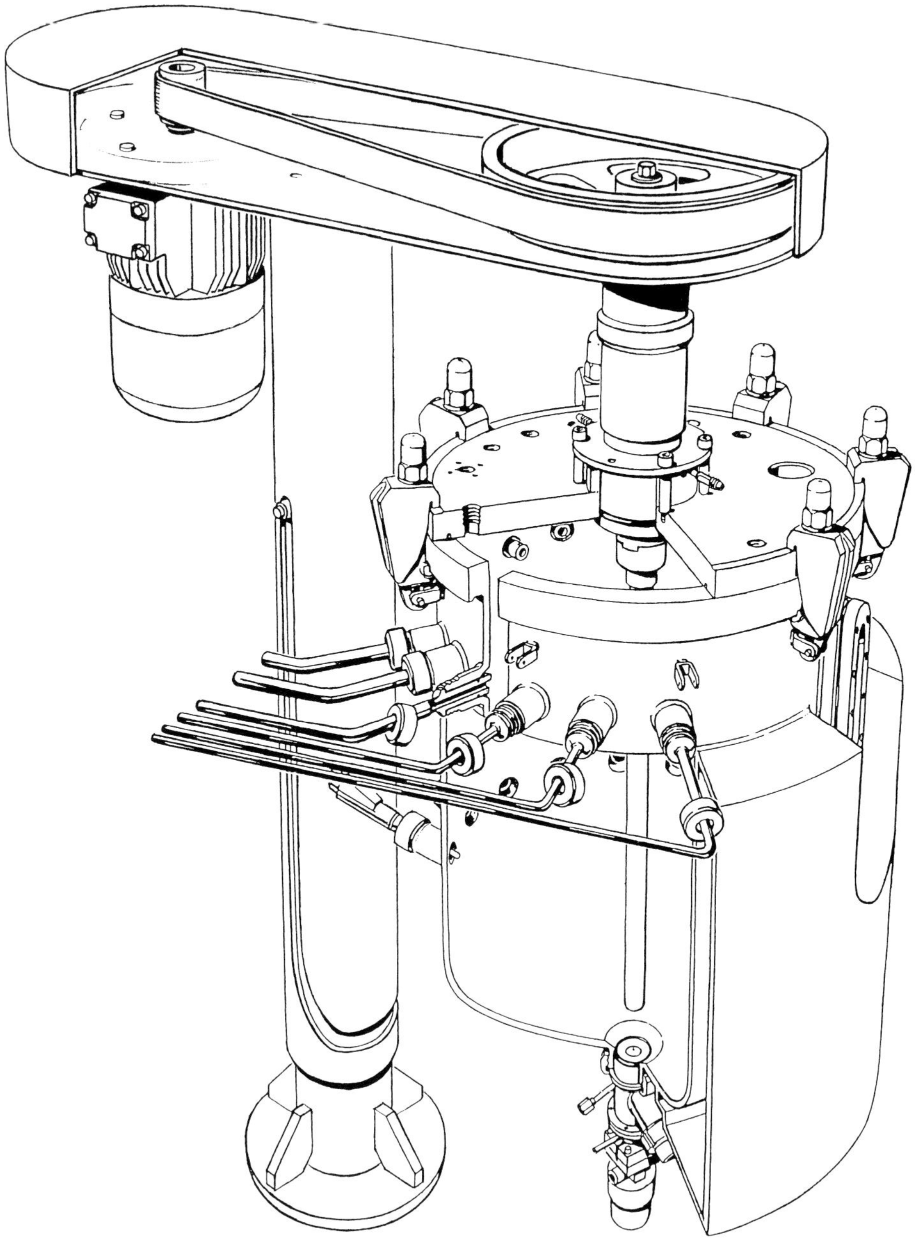

Example of a top-mounted drive combined with a self-positioning lifting column.

a Disk turbine
b Hollow blade stirrer
c,d Pitched blade impeller

e Intermig impeller
f Pitched paddle impeller
g Propeller
h Pitched blade propeller

i Helical ribbon impeller
j-l Anchor impellers

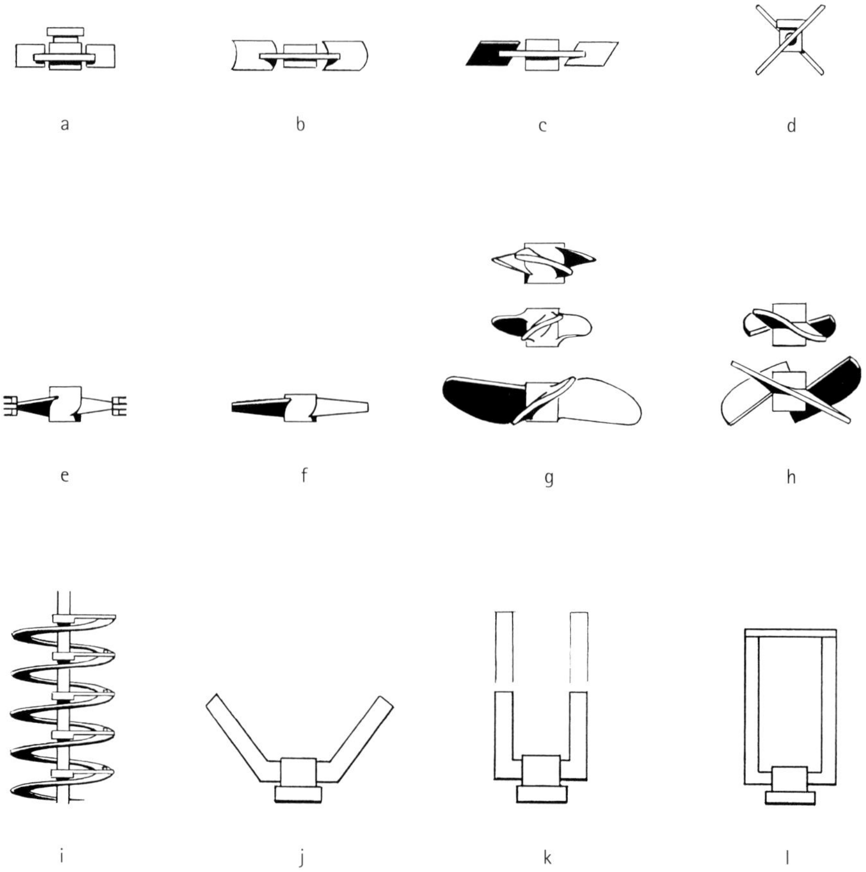

The basic tasks of the impeller include mixing of the medium, feeding oxygen into the medium, and optimizing heat transmission. Processes vary greatly in their stirring requirements: Some need very intensive stirring movements with high power input (e.g. microbial cultures), while others depend on gentle stirring (e.g. cell cultures). A vast array of different impeller types is available for the key stirring tasks such as homogenizing, dispersing, suspending, etc. Turbines (flat-bladed Rushton) and propellers (marine impellers) are widely used in fermentation processes.

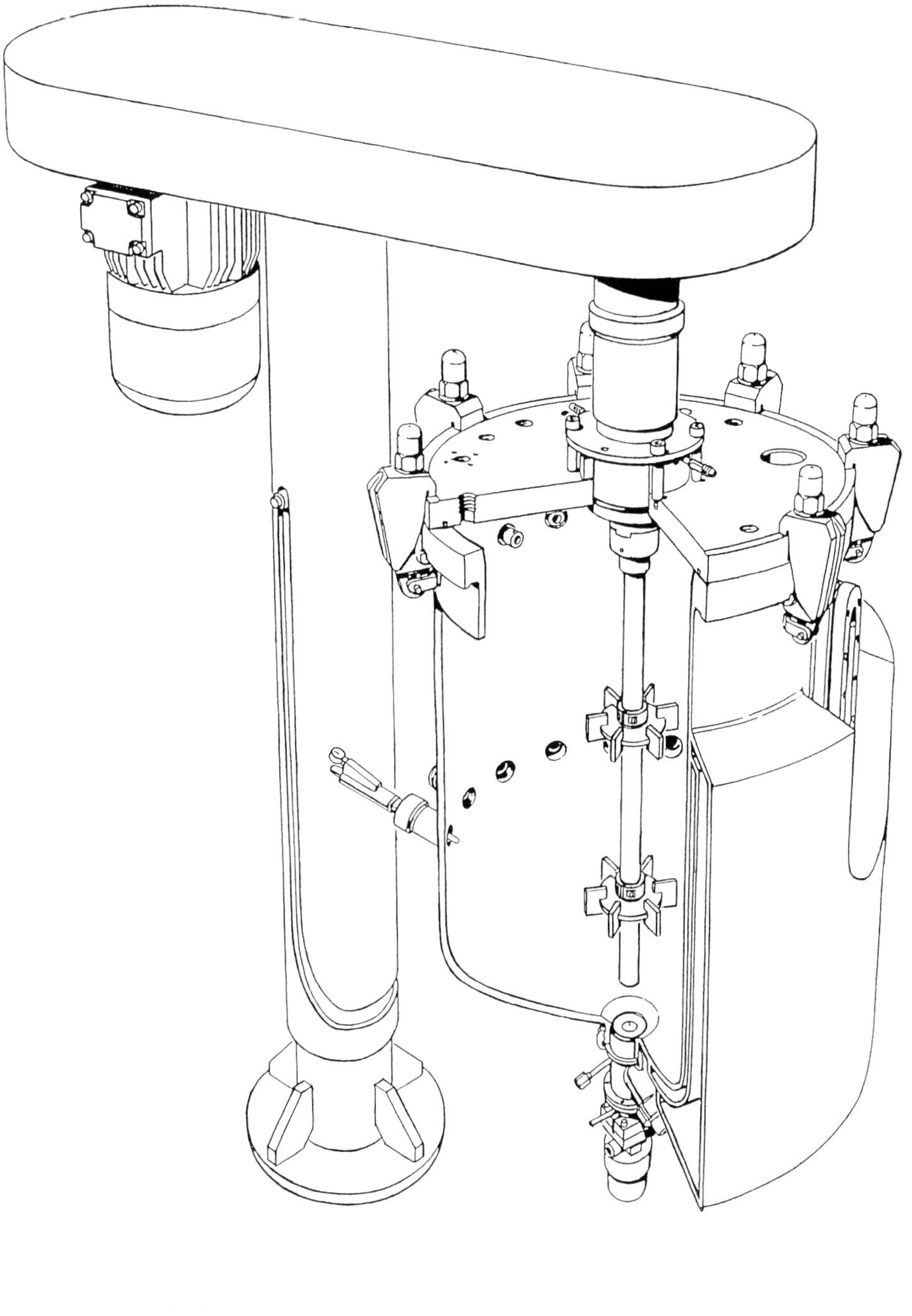

A stirrer shaft equipment with disk turbine.

The disk turbine with a radial primary flow direction is used predominantly where turbulent flow is needed. It is especially suitable for dispersing liquid / liquid substances and to achieve a high phase-related exchange surface which determines the rate of substance transport.

The propeller with an axial primary flow direction involves only very small shear forces. This leads to very gentle but nevertheless thorough mixing. For this reason, the propeller is recommended especially for the cultivation of animal cells. But it is also an excellent tool for homogenizing and suspending liquids a well as dispersing solids in liquids.

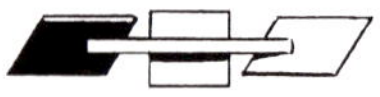

Pitched blade impellers are particularly suitable for homogenizing, suspending, and dispersing liquid / liquid and solid / liquid substances, as well as the gentle stirring of media.

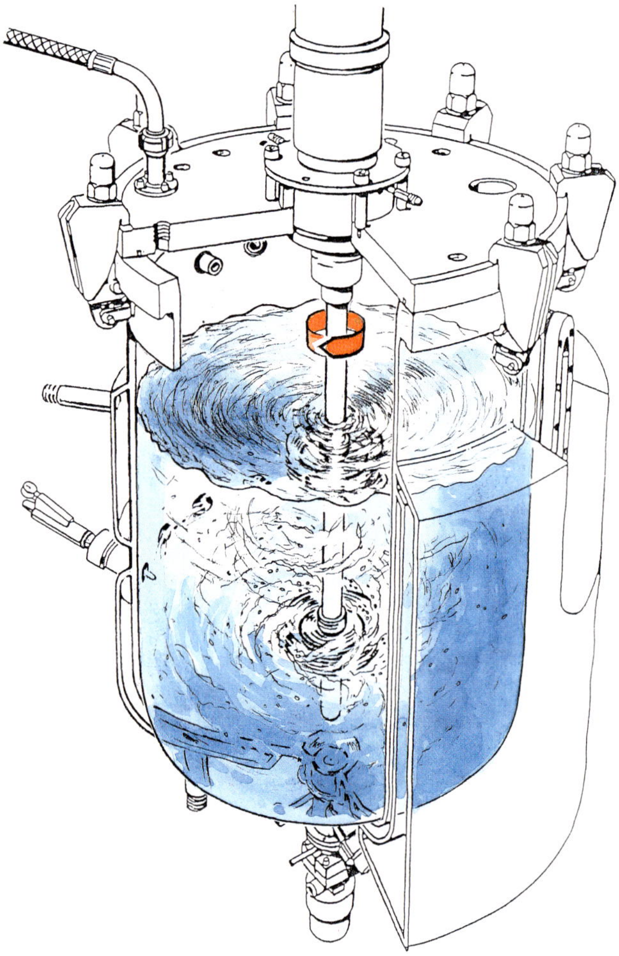

Whether the stirring task is properly performed depends above all on the flow pattern. Every impeller is characterized by the typical flow pattern it generates. It is used wherever the specific flow pattern achieves the optimal mixing effect for the corresponding process.

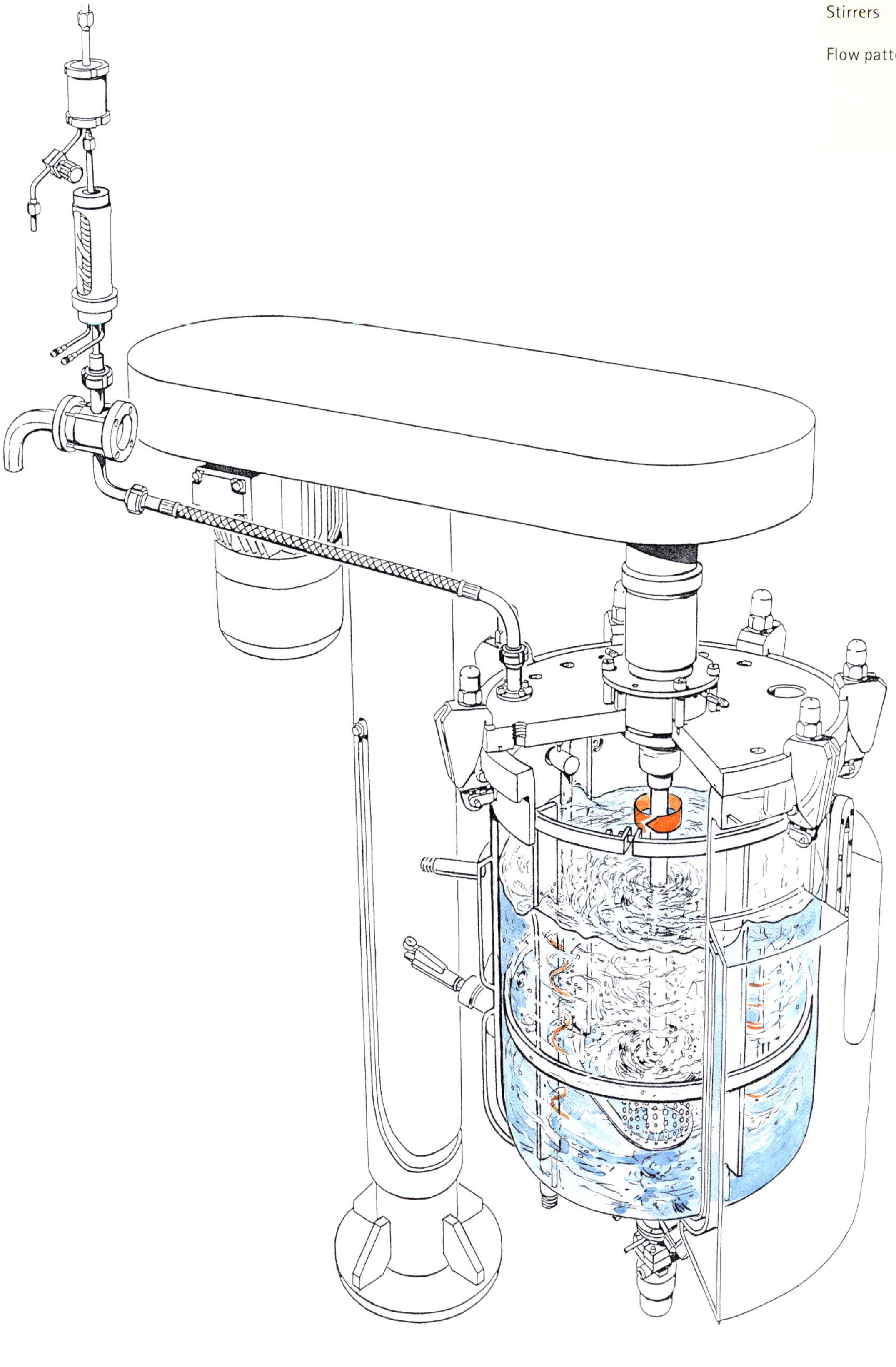

Interior view of a stirrer-equipped reactor. Built-in components may sometimes influence the flow pattern. In case of very intensive stirring tasks, care should be taken that all interior components are mounted safely.

Baffles designed as als entry / exit aid

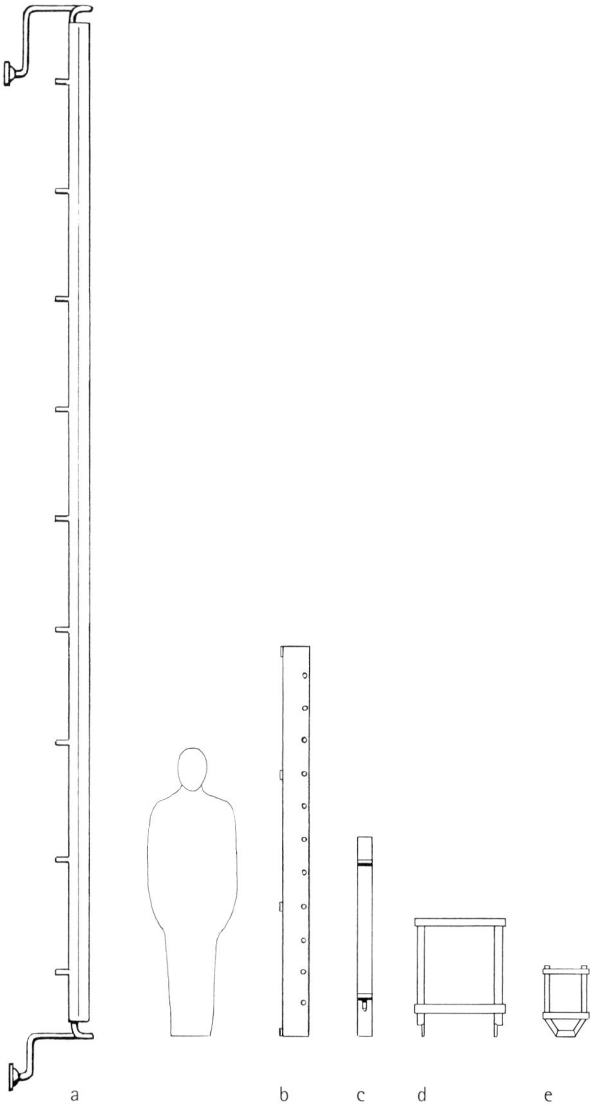

a Perfused baffle
b Baffle with visual level indicator
c Removable baffle
d Baffle basket
e Baffle basket with bayonet catch

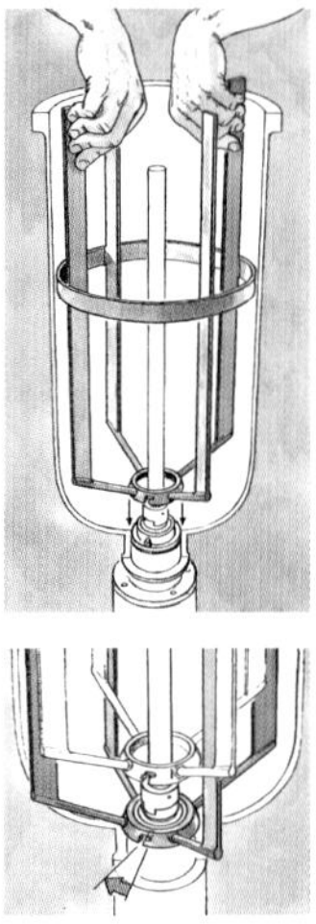

Baffle basket fastened with bayonet catch

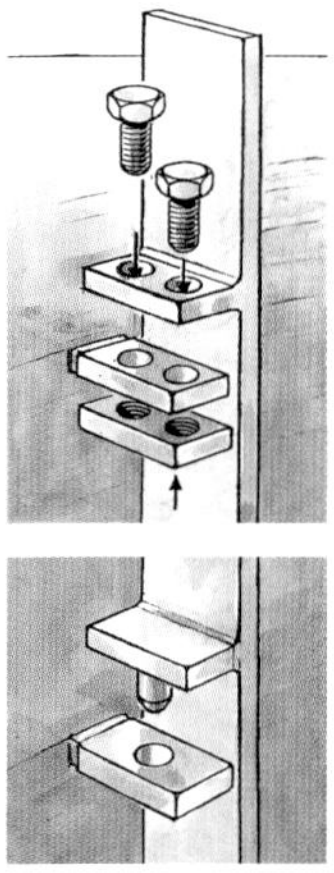

In case of high power inputs, reliable fastening of the baffles is essential

The task of a baffle can be described as follows: Its purpose is to prevent rotary flow and the formation of thrombi caused by rotary drives. With a baffle, the stirrers in the reactor generate flow patterns that positively influence oxygenation, heat transfer, and mixing properties. Baffles may be permanently installed or removable and can perform various additional tasks such as: perfused heating / cooling elements, visual level indication, carriers for special attachments, or even entry / exit aids. Permanently mounted baffles have the disdvantage that they cannot be disassembled when necessary. If they are incorrectly welded in, there is a risk of breakage of the reactor wall. A simple way to avoid using baffles is to mount the stirrer eccentrically.

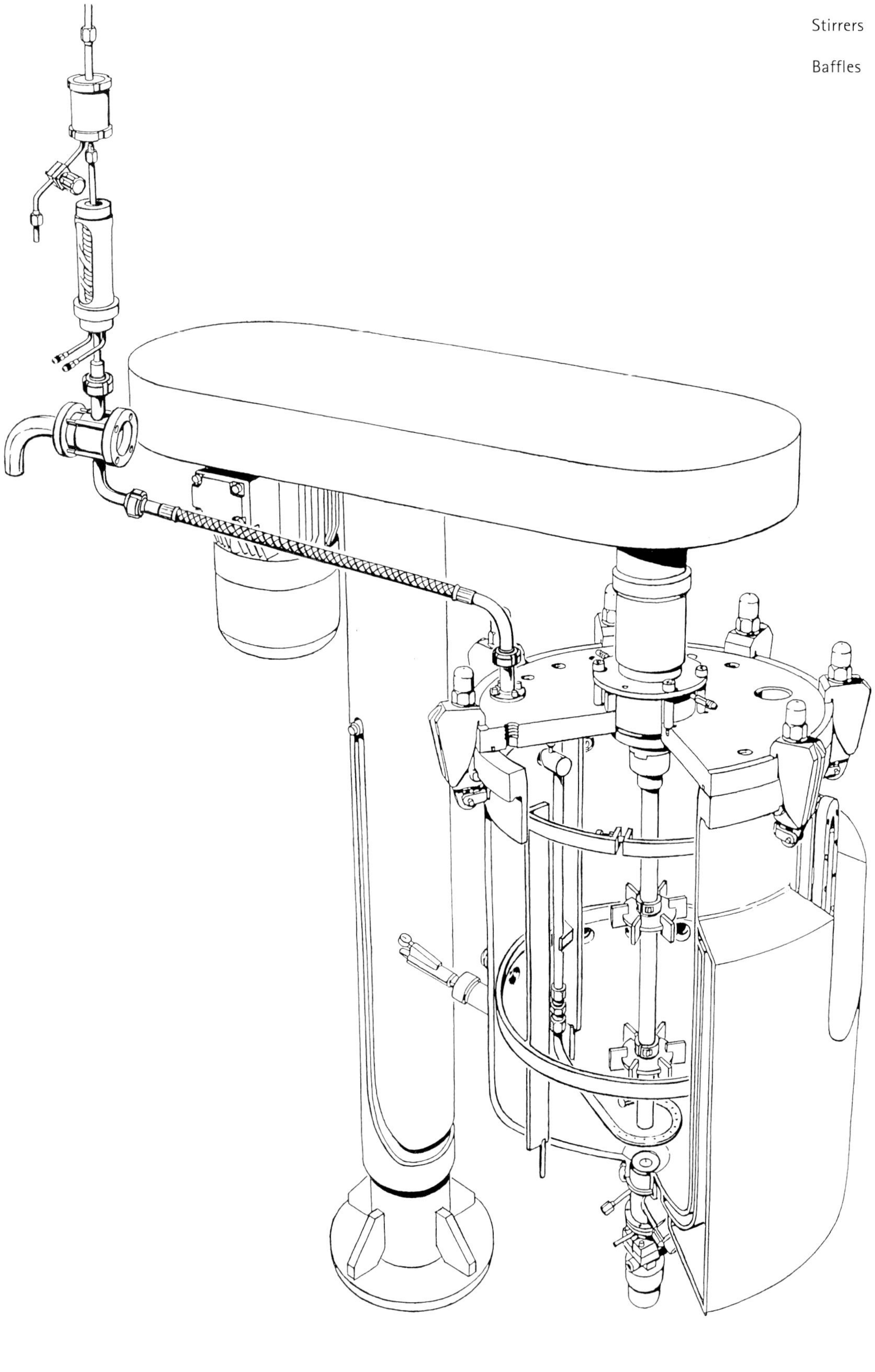

Baffle baskets have the advantage that they are flexible with regard to their position. This is particularly useful when many attachments or retrofittings are mounted.

Non-return valve

Pressure gauge

Flowmeter

Pressure-reducing valve

Fermentation processes require a supply of air to the medium. In principle, the air is preconditioned inside the air feed line so that the required quantity reaches the medium in sterilized form. Air supply lines may be designed as simple air feeds or involve multiple gas-mixing units with manual or automatic control. Like many other product lines, the air supply section must be designed to allow steam sterilization. Since filters are sensitive to humidity, care should be taken that they cannot be clogged by condensate formation.

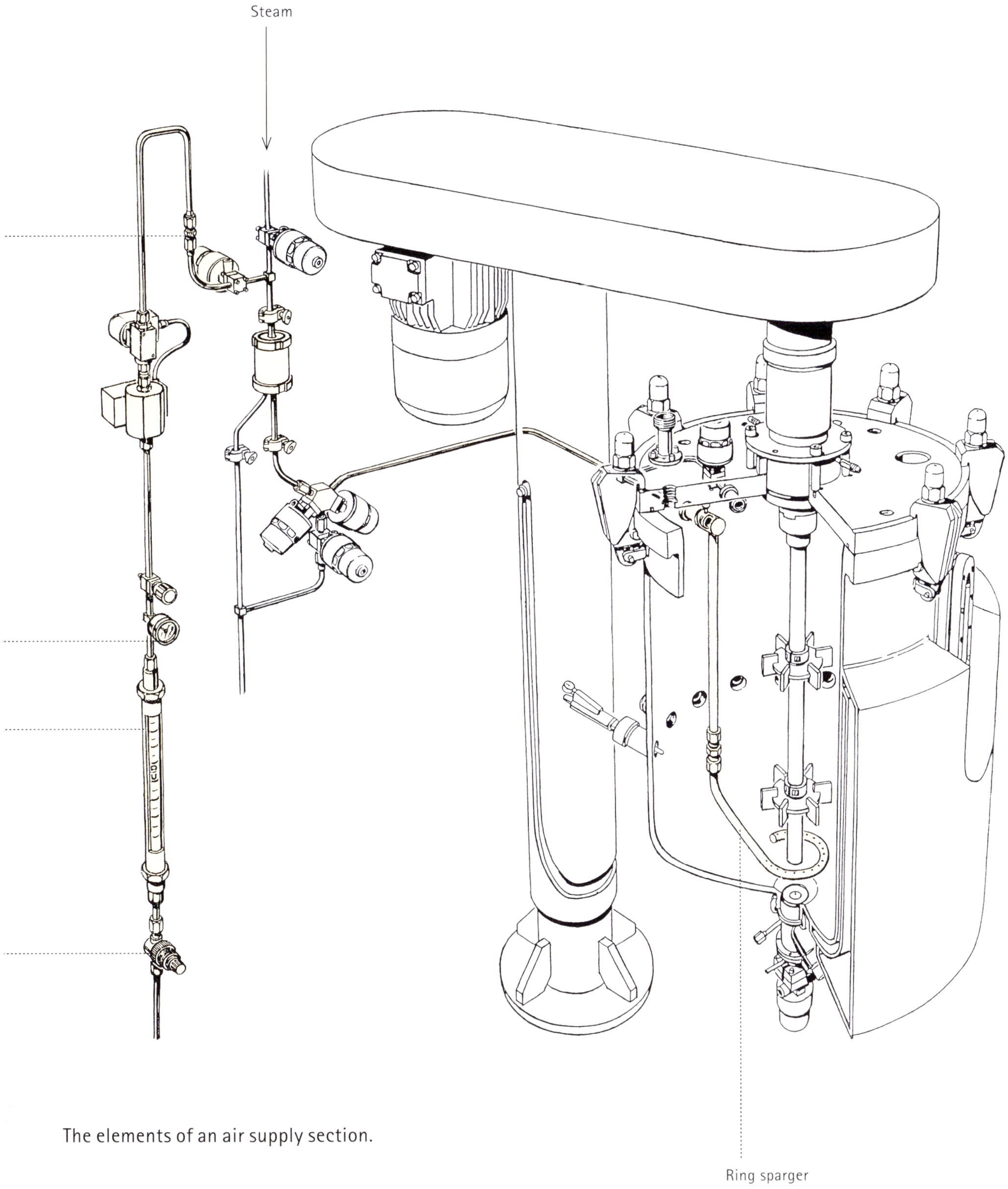

The elements of an air supply section.

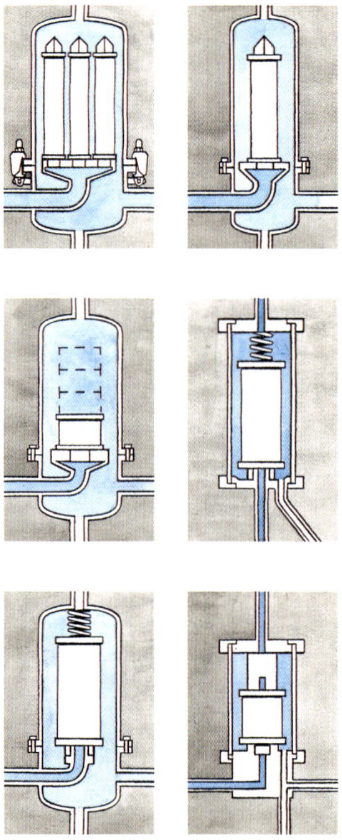

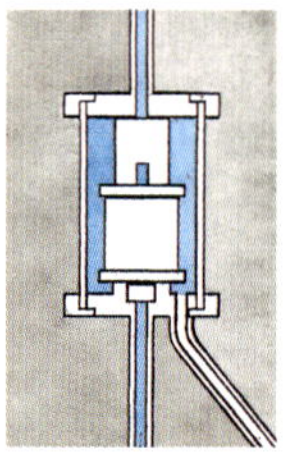

Overview of the various designs and sizes of air filters

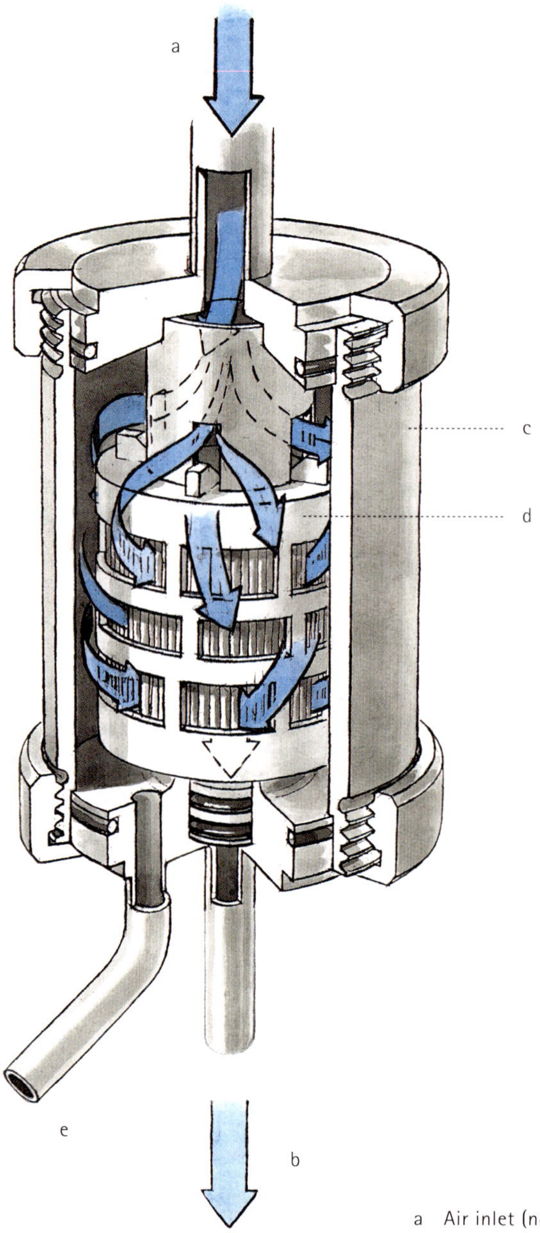

a Air inlet (non-sterile side) / steam inlet
b Air inlet (sterile side)
c Filter housing
d Filter cartridge
e Condensate outlet

Filter housings and air filter cartridges are available in a multitude of designs and dimensions. Which type of air filter is selected for a particular process depends on the most diverse factors: type of filtration (particle, micro-, or ultrafiltration), medium (gas or liquid), and required filter surface (aeration rate). With respect to design, there are various types of housing connections (clamp, flange) or adapter designs. The housing shapes are adapted to the filter cartridges. Caution is indicated when selecting adapters because not all renowned manufacturers produce filter cartridges that are compatible among each other. Because filter cartridges are replacement elements, good accessibility and easy disassembly is an important consideration when installing filter housings.

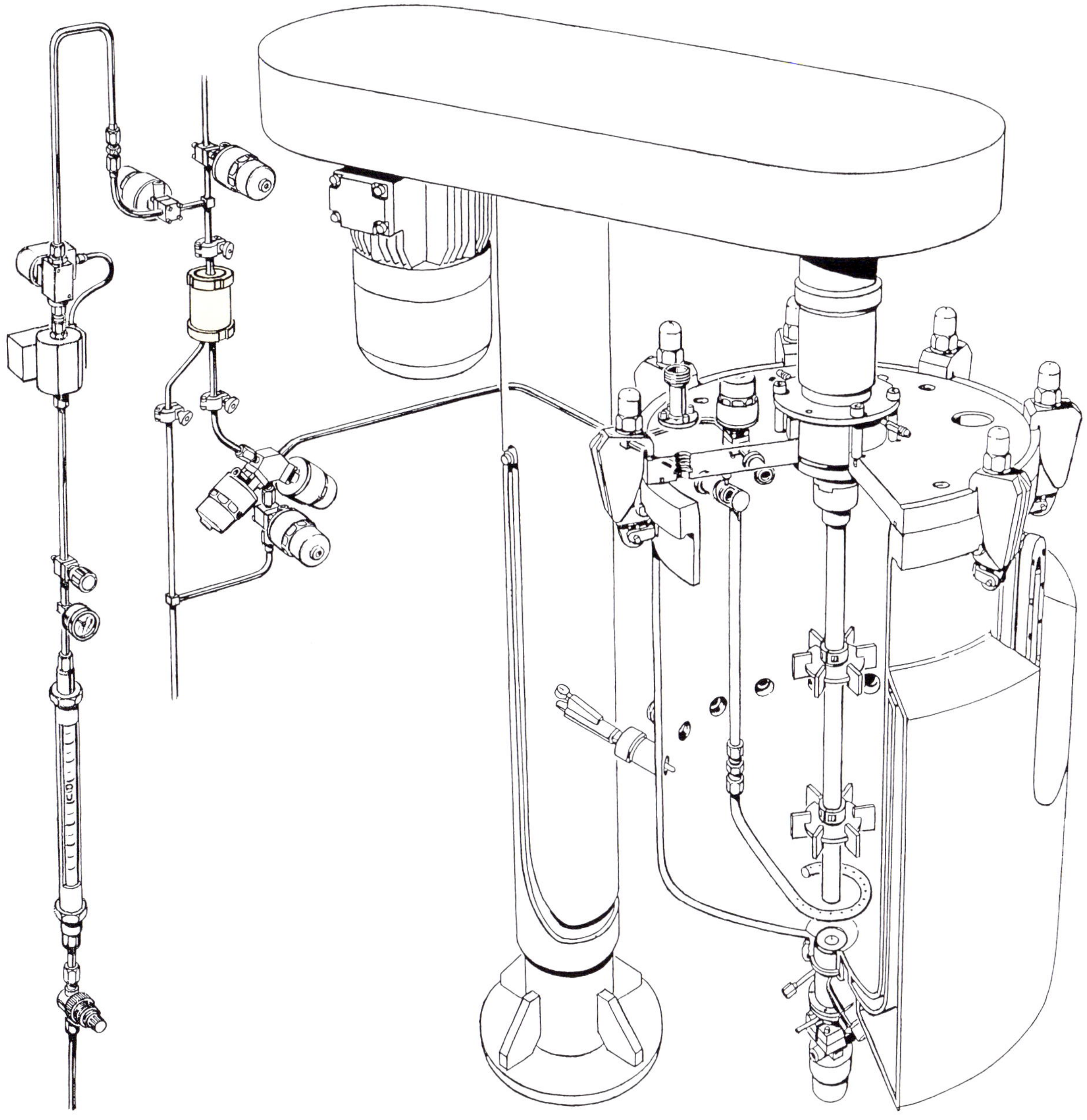

Installation details of an air inlet filter.

a Ring sparger
b Sintered pipe
c Aeration immersion tube
d Ring sparger
(through lower port)
e Sintered pipe
(through lower port)

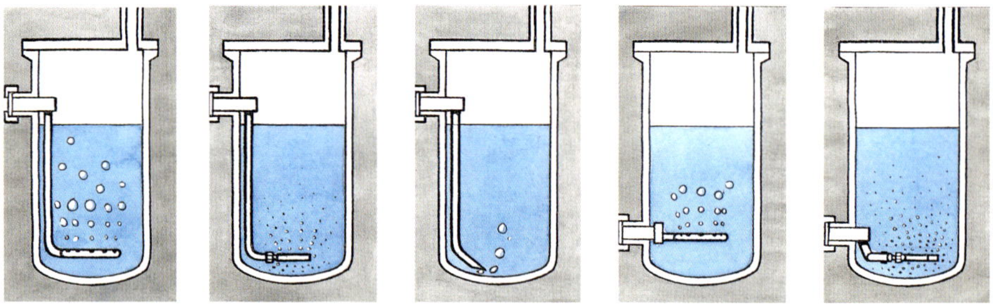

a b c d e

f Headroom aeration

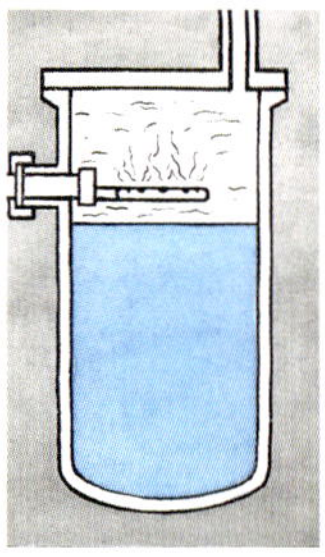

f

g Aeration tube (helical)
h Axial aeration tubes
(rotating)
i Axial aeration tubes
(tumbling)

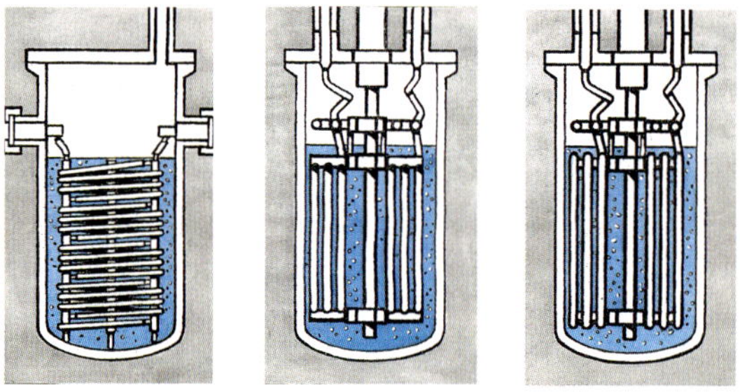

g h i

Many processes require aeration of the medium. Aeration systems and types are available in various designs. The choice of system depends on the aeration rate and the medium and includes: ring spargers with aeration holes to generate many small, homogeneously distributed bubbles, sintered pipes to generate minuscule bubbles and to protect sensitive media, bubble-free aeration systems for oxygenation by diffusion.

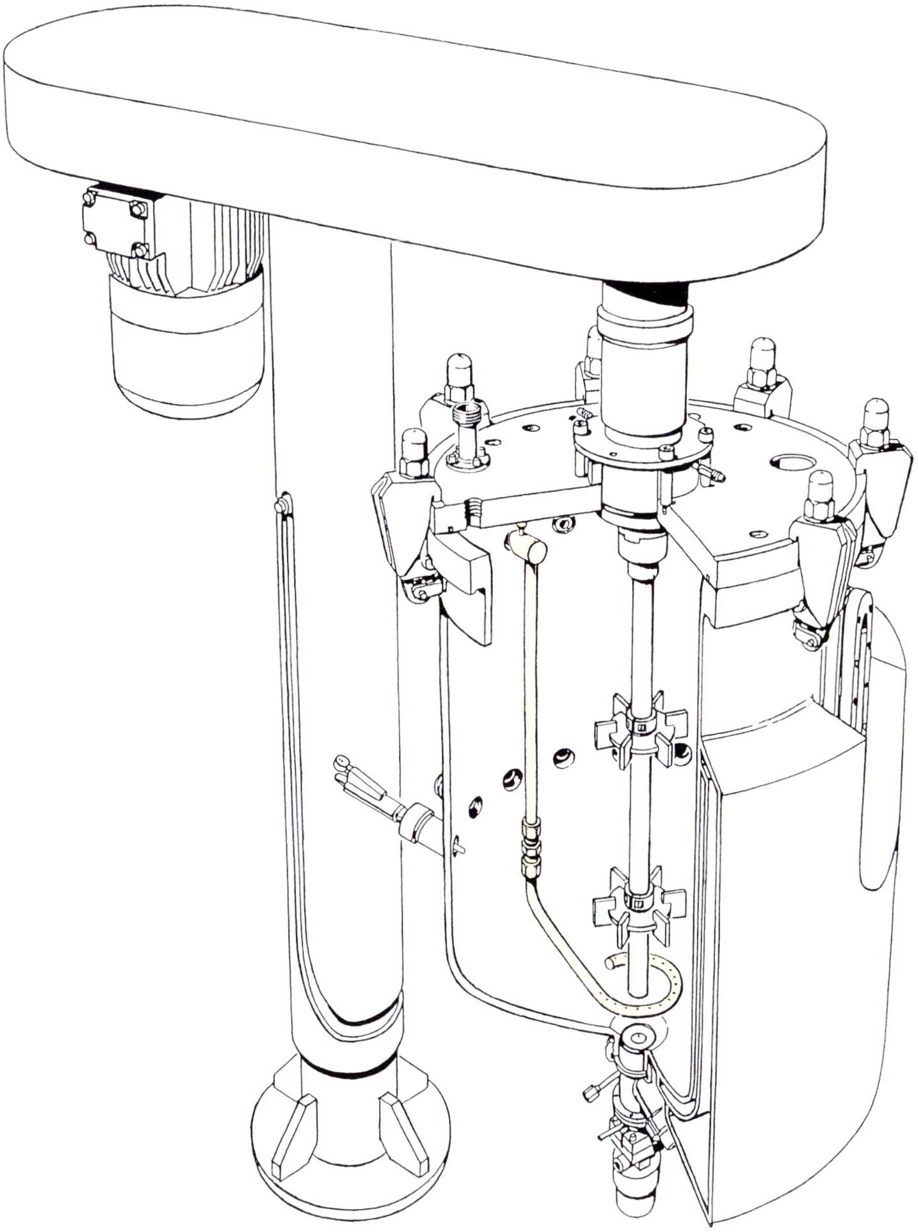

A fermenter configuration with a ring sparger.

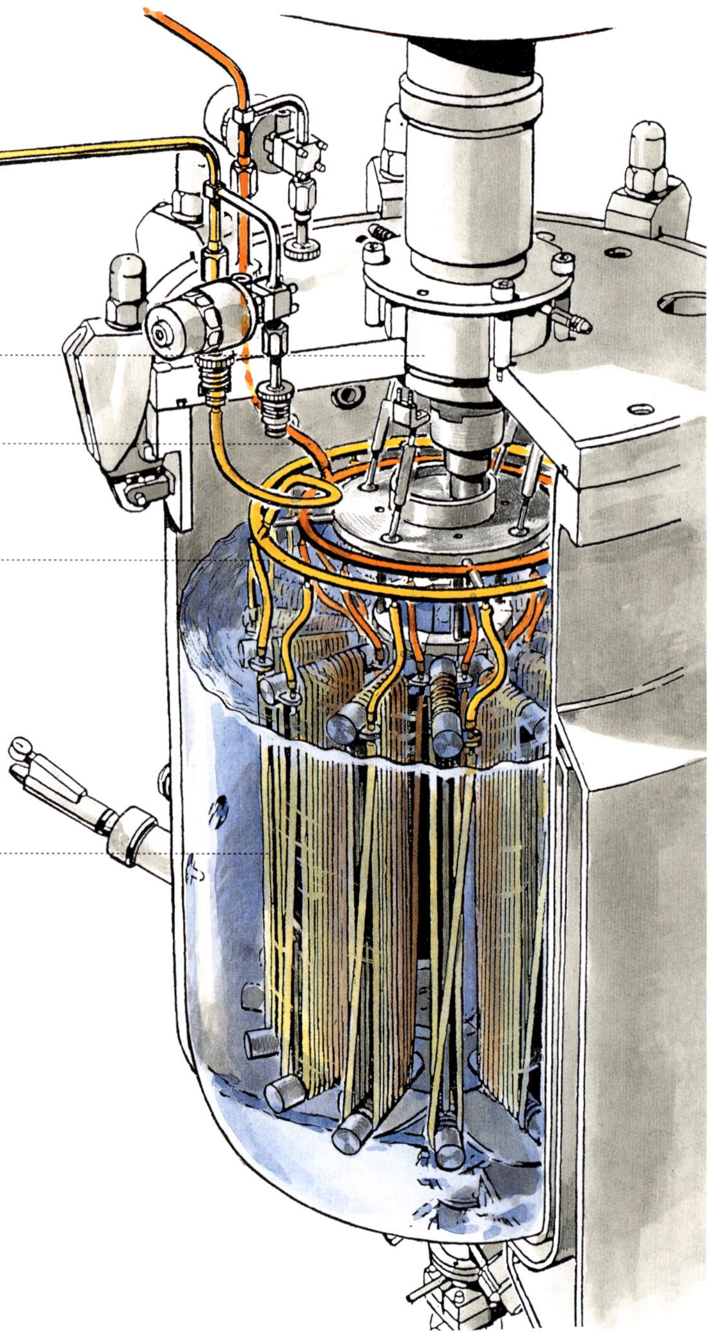

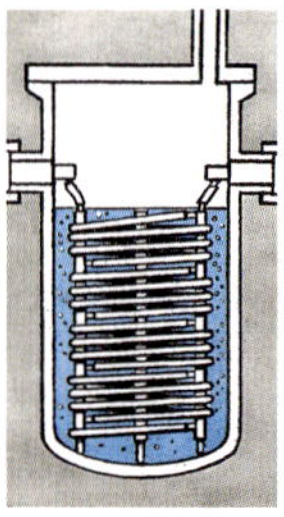

Radially wound tubes, static

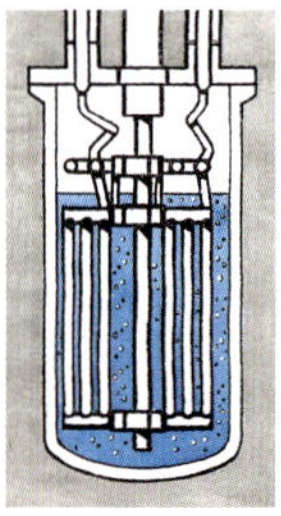

Axially arranged tubes, rotary movement

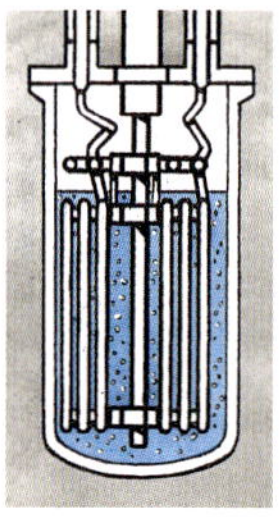

Axially wound tubes, tumbling movement

There are many different ways to aerate a reactor. In the case of shear-sensitive cell cultures, gentle stirrer movements and aeration are particularly important in order to avoid mechanical destruction of the cells. A well established method in practice is bubble-free aeration. With bubble-free aeration, tubes immersed in the medium are perfused with sterile supply air. The air permeates through the tube wall and enters the medium. Since this process often requires a very large exchange surface, there is usually no space for stirrers. In such cases, the aeration system is designed in such a way that it generates a stirring movement. The large figure shows an example of tube aeration with integrated tumbling movement.

Supply air

Bubble-free aeration

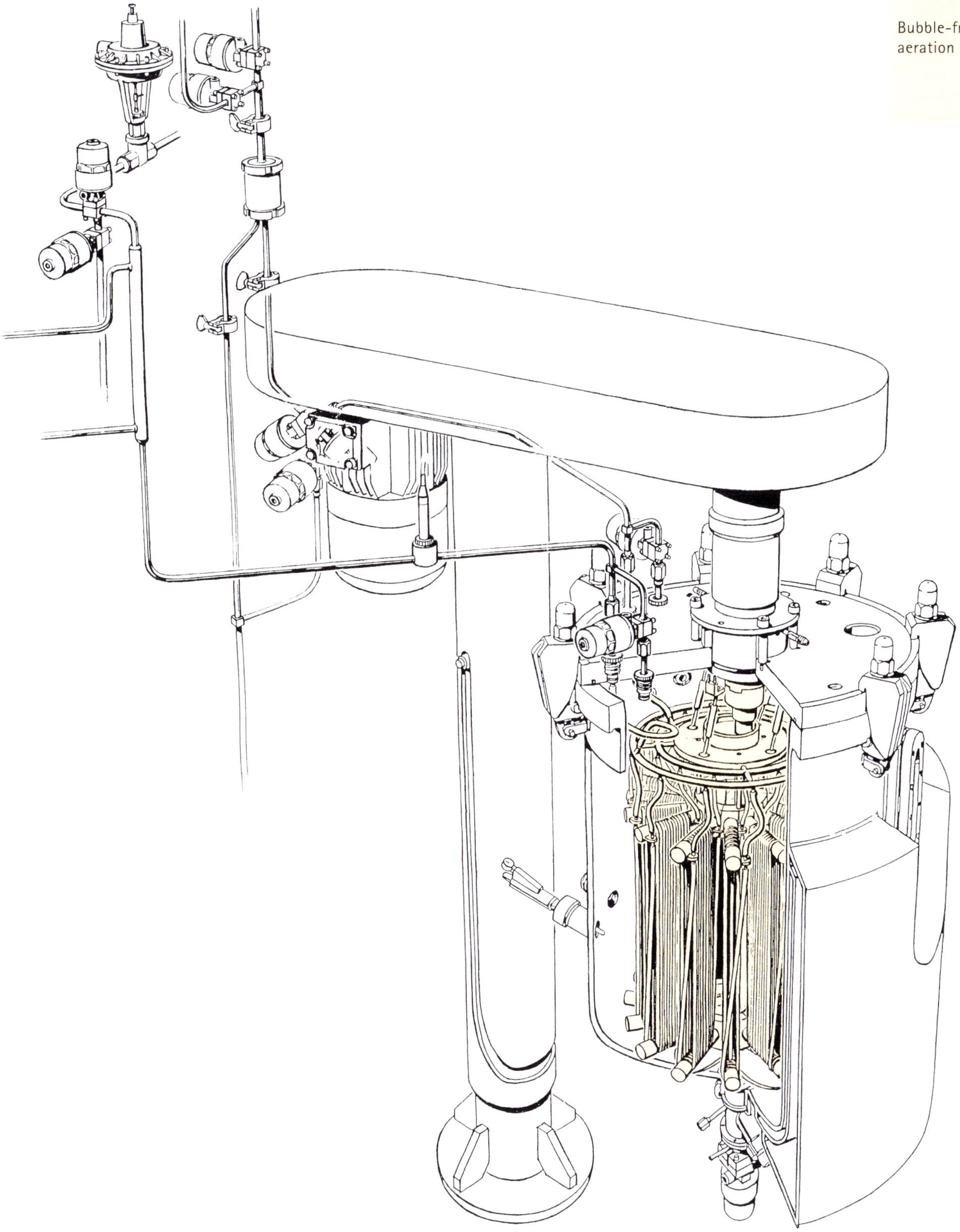

Bioreactor with integrated tube aeration basket. In addition to the axially arranged tubes, the figure also shows a drive that generates a tumbling movement.

Exhaust air from the reactor is discharged through the exhaust air line. The design of the exhaust line may depend on a number of different criteria. An exhaust air filter ensures that only filtered air can escape. To minimize the loss of liquid, cooling elements (reflux coolers) may be used which condensate the exhaust air and route the liquid back to into the medium. In turn, this requires low flow rates in the piping. Like the supply air section, the exhaust air section should also be sterilizable and controllable in the manual or automatic modes. An important consideration in designing the exhaust air line is the prevention of vacuum formation in the reactor as a result of cooling processes, for instance after sterilization.

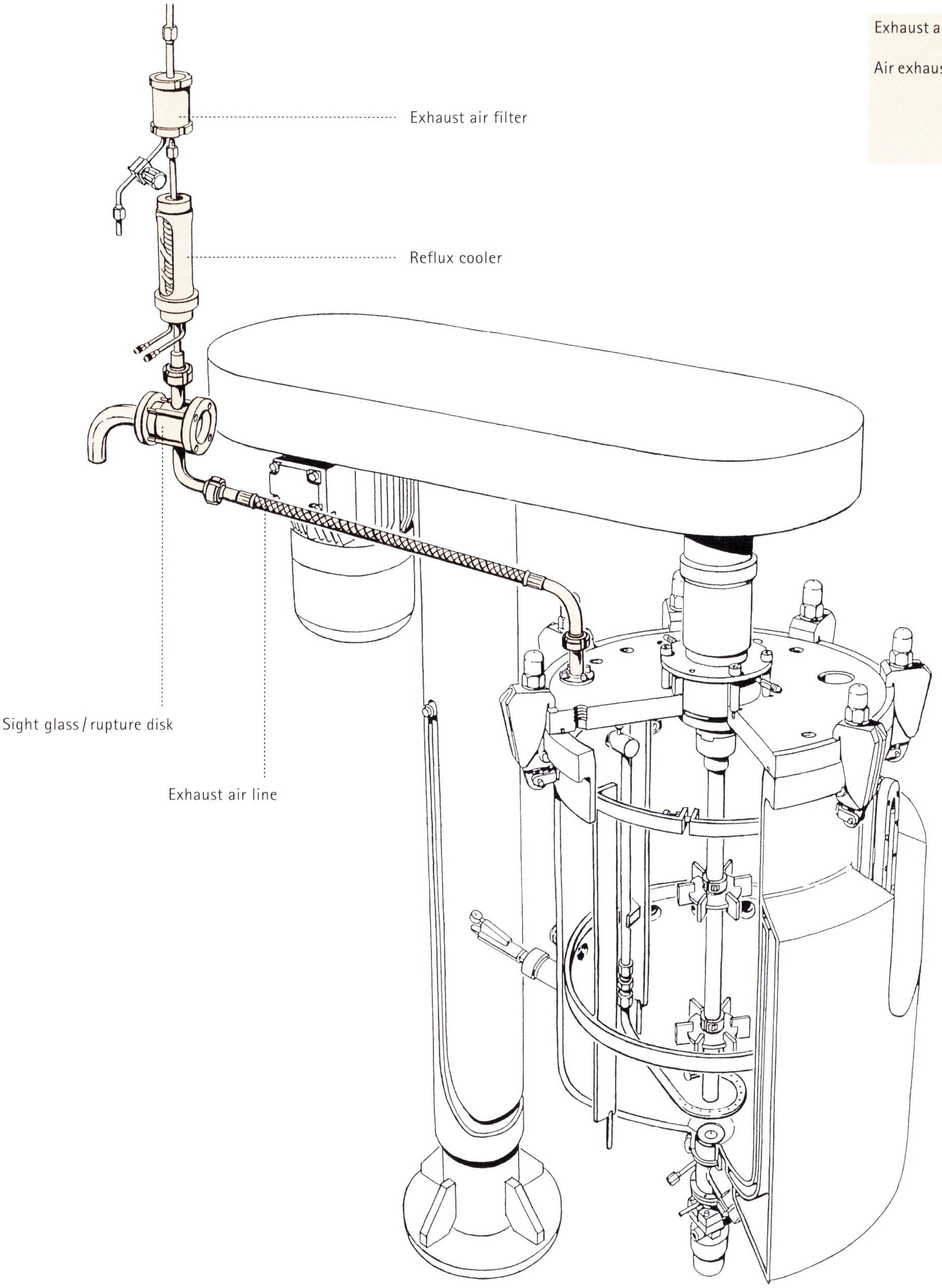

The elements of an air exhaust section.

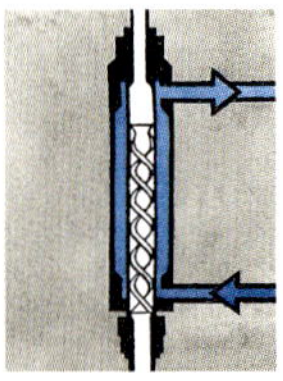

Baffle

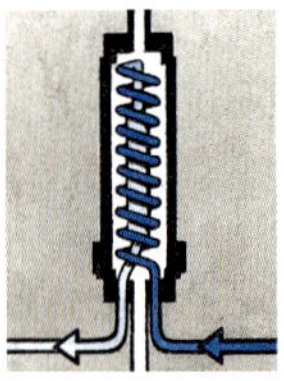

Spiral tube

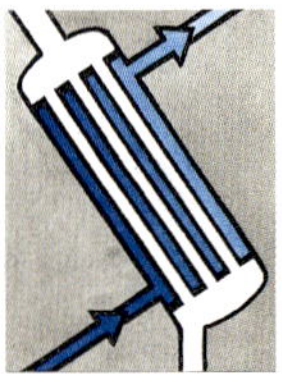

Tube bundle

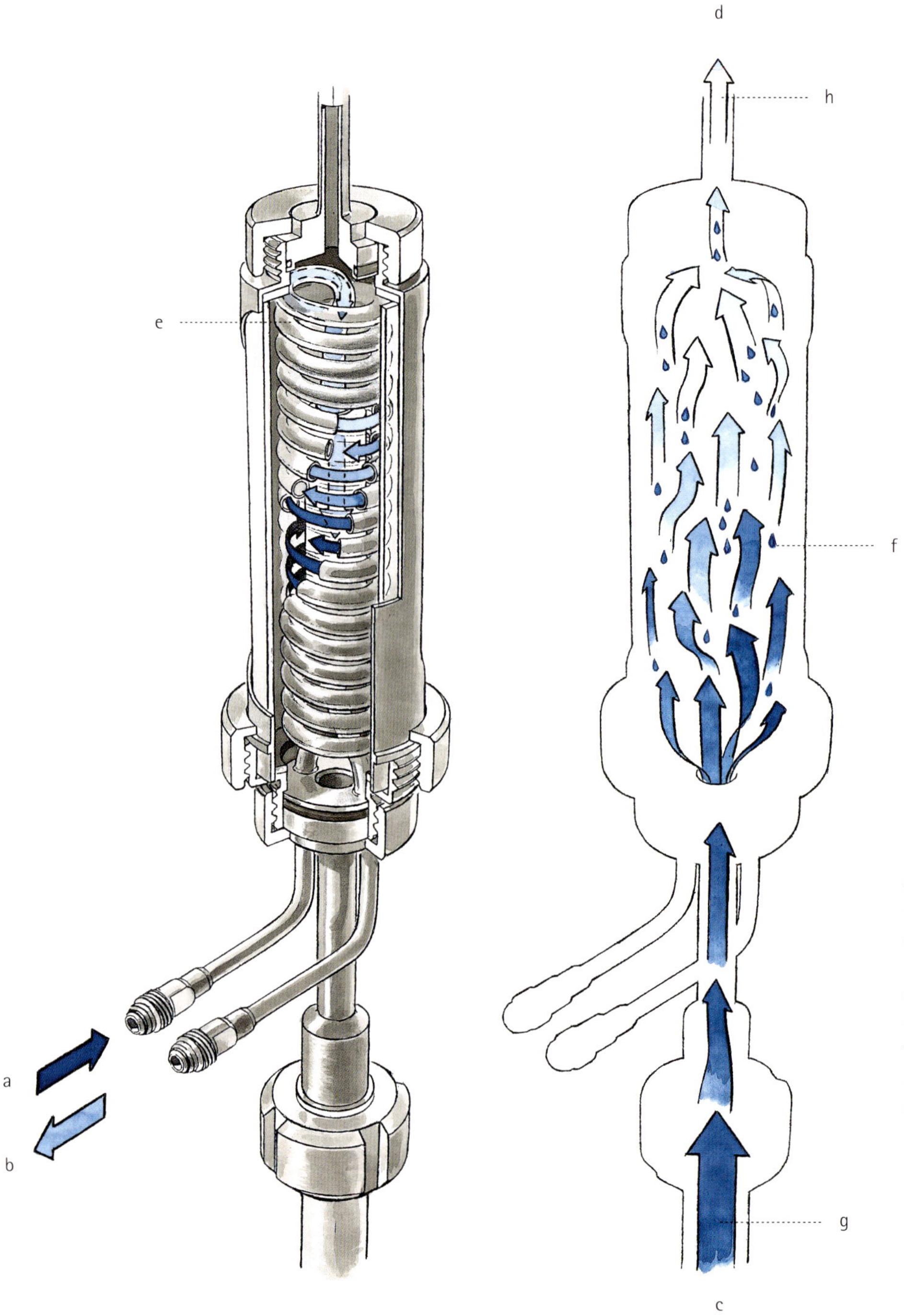

a Coolant inlet
b Coolant outlet
c Exhaust air inlet, condensate return
d Exhaust air outlet
e Cooler with spiral tube
f Condensate
g Saturated exhaust air
h Unsaturated exhaust air

The air escaping from the fermenter has been heated and is therefore saturated with humidity. This leads to a loss of liquid, which is undesirable in many processes. This loss can be greatly reduced by using a reflux cooler. In a cooler, the escaping exhaust air is cooled down until the humidity condensates and flows back into the reactor. It is essential that the design of the unit as well as the dimensions of the piping and the layout of the return pipes allow a trouble-free reflux of the medium. The flow rate of the exhaust air must be controlled accordingly.

Exhaust air

Reflux cooler

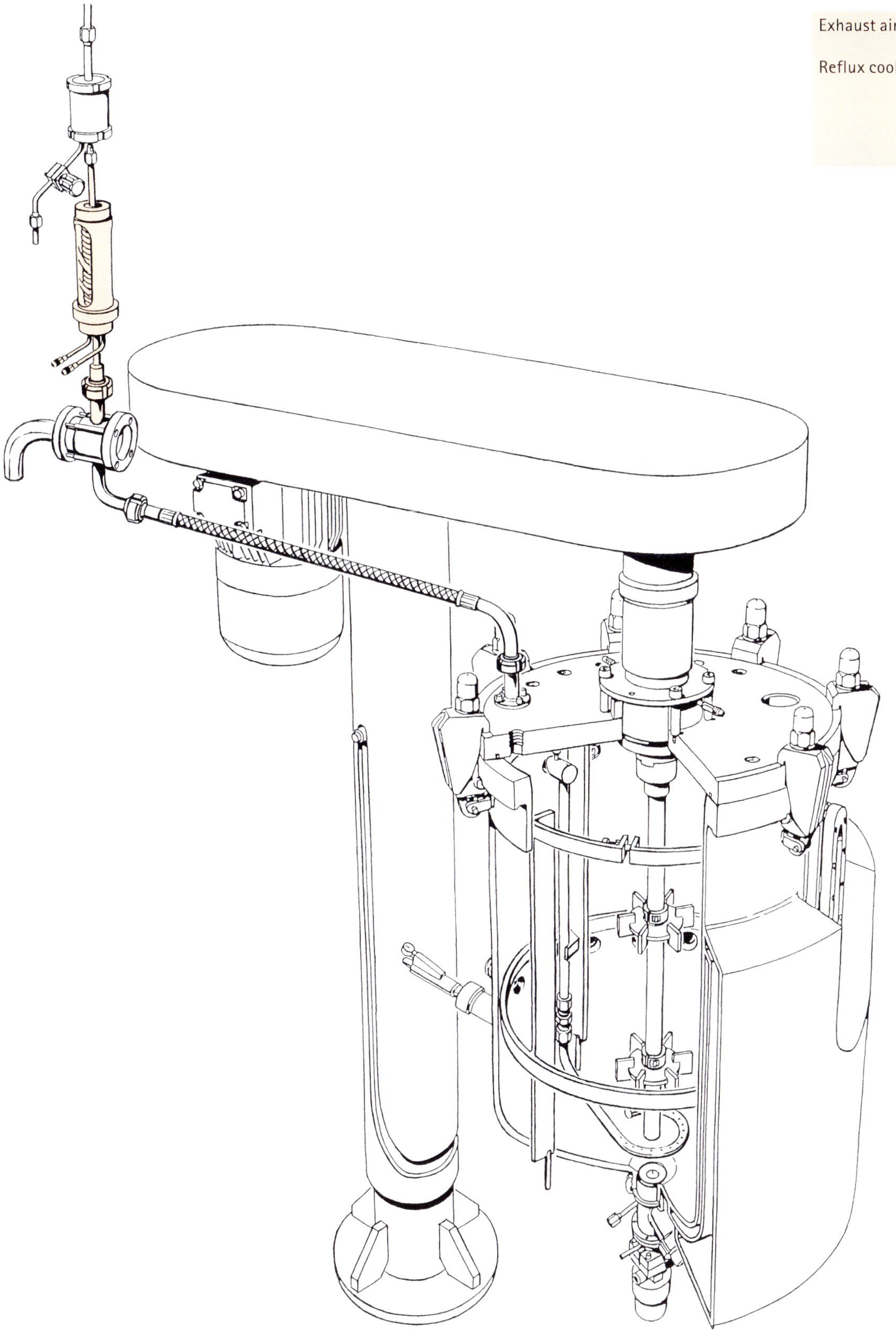

Bioengineering reflux cooler.

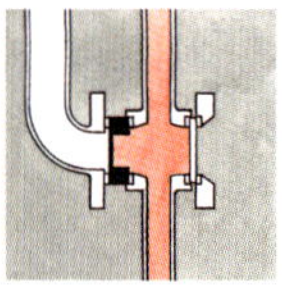

Pipe-mounted sight glass / rupture disk fitting

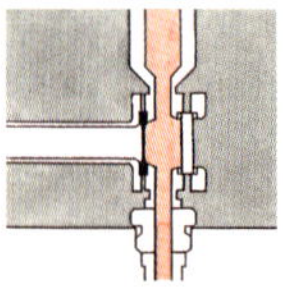

Lid-mounted sight glass / rupture disk fitting

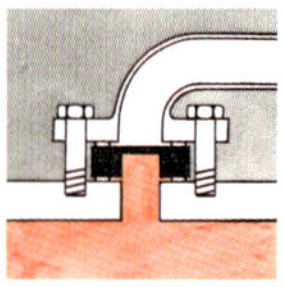

Lid-mounted graphite rupture disk fitting

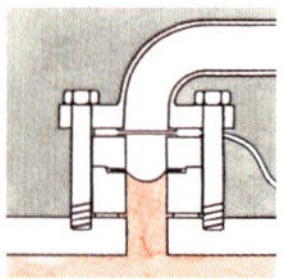

Lid-mounted steel rupture disk fitting

A sight glass / rupture disk fitting mounted into the exhaust line is perfused during the process. The overpressure is within the maximum permissible service pressure.

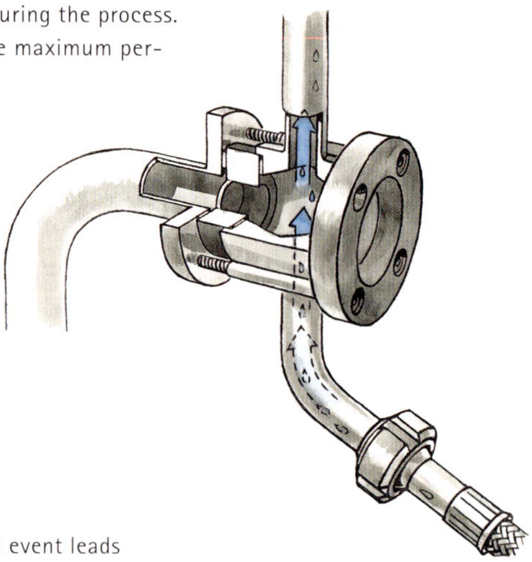

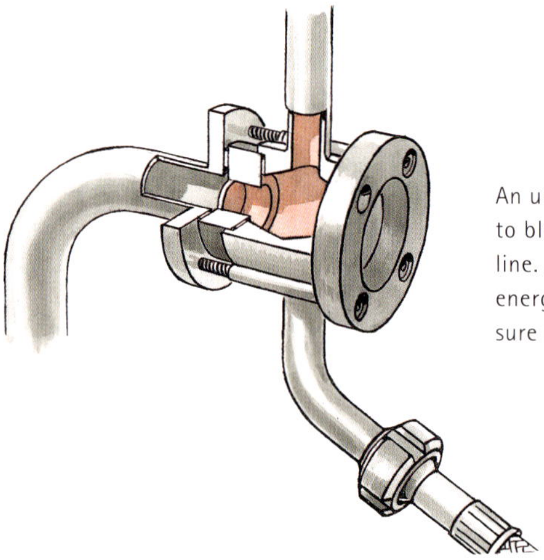

An unforeseen event leads to blockage of the exhaust line. As the supply of energy continues, the pressure rises steadily.

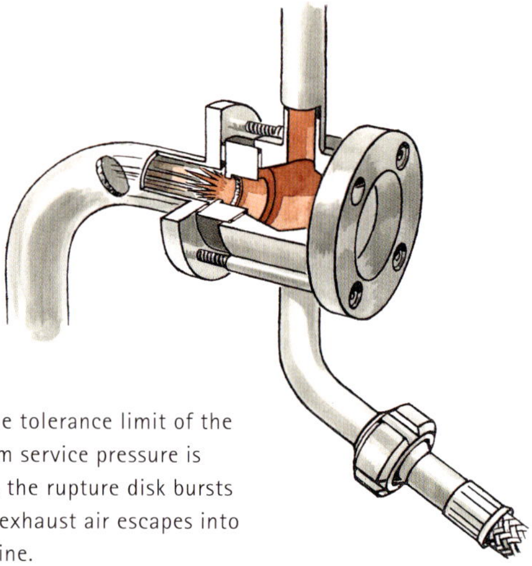

When the tolerance limit of the maximum service pressure is reached, the rupture disk bursts and the exhaust air escapes into an exit line.

Every pressure device should be protected by a safety element to ensure that the permissible service pressure is never exceeded. In most cases, this is achieved with a safety valve which is accurately set and sealed by a calibration authority. However, such safety valves may incorporate undesirable dead spaces and are therefore poorly suited for sterile applications. An alternative to safety valves are rupture disks which are available in a great variety of designs and materials. A far better solution is to make rupture disks unnecessary by designing and dimensioning all pressurized compartments in such a way that they are safeguarded against overpressure.

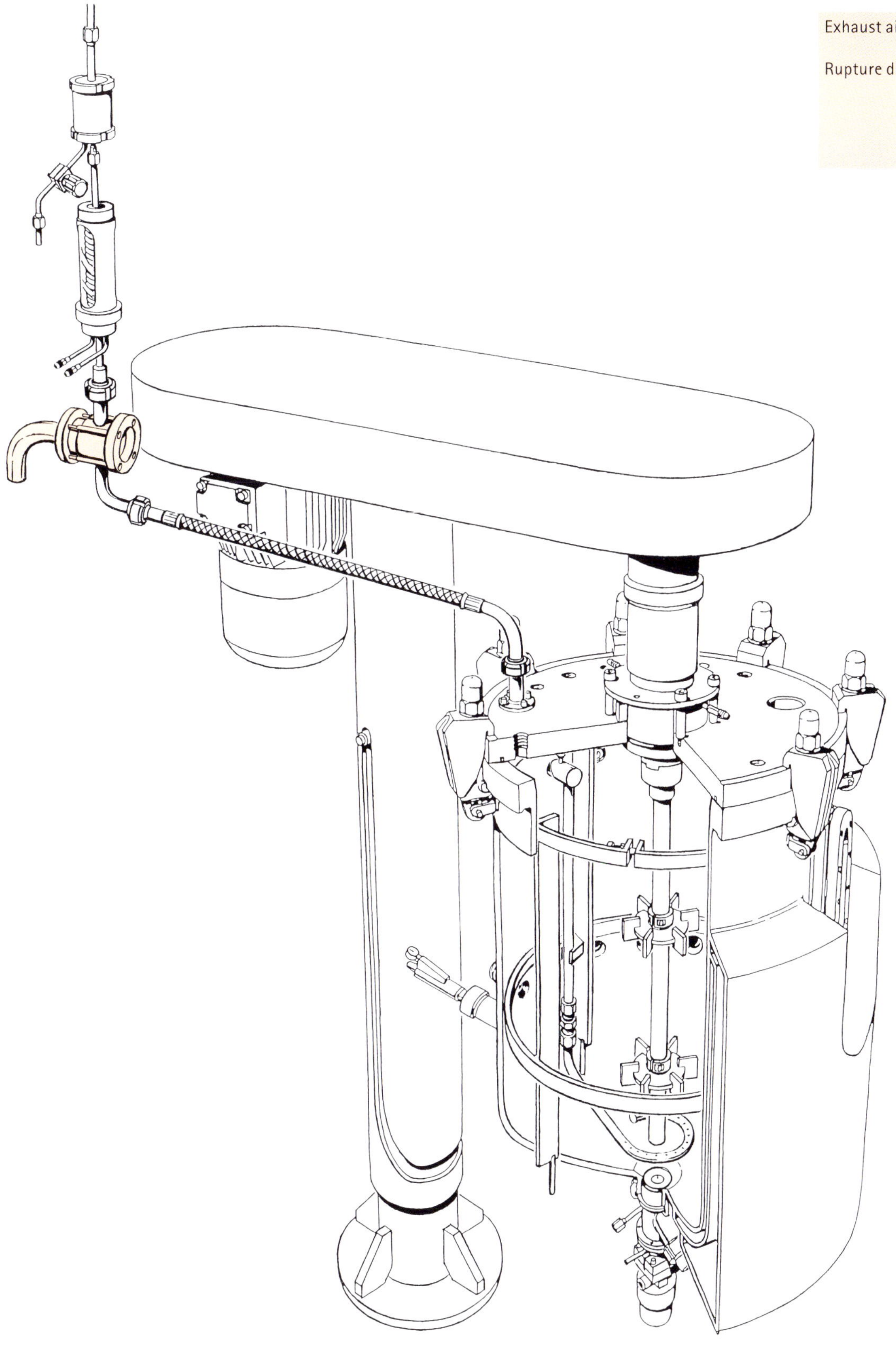

Example of a sight glass / rupture disk fitting mounted into an exhaust line.

Bottom drain valve open

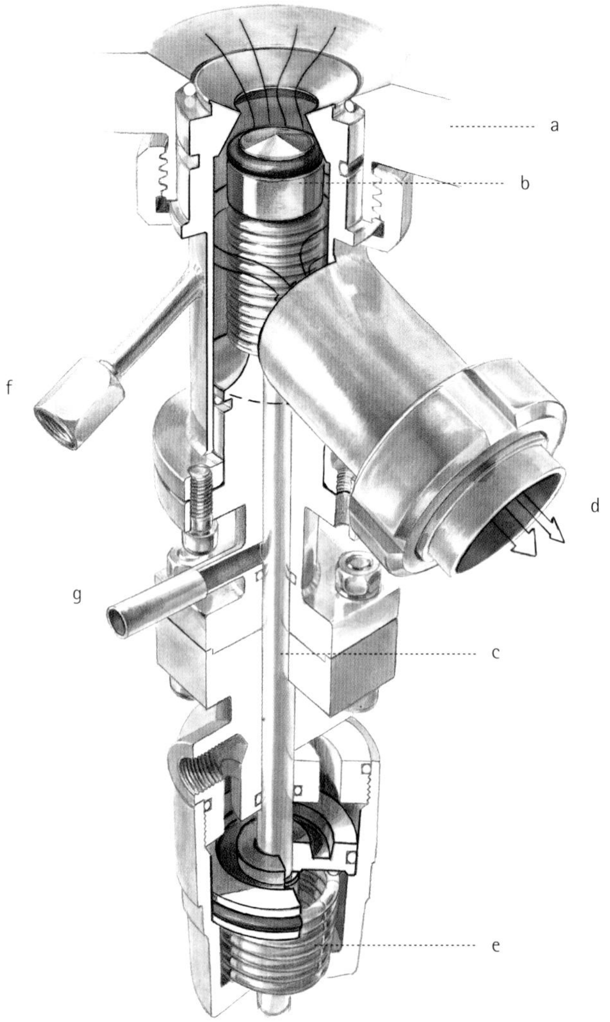

Bottom drain valve closed

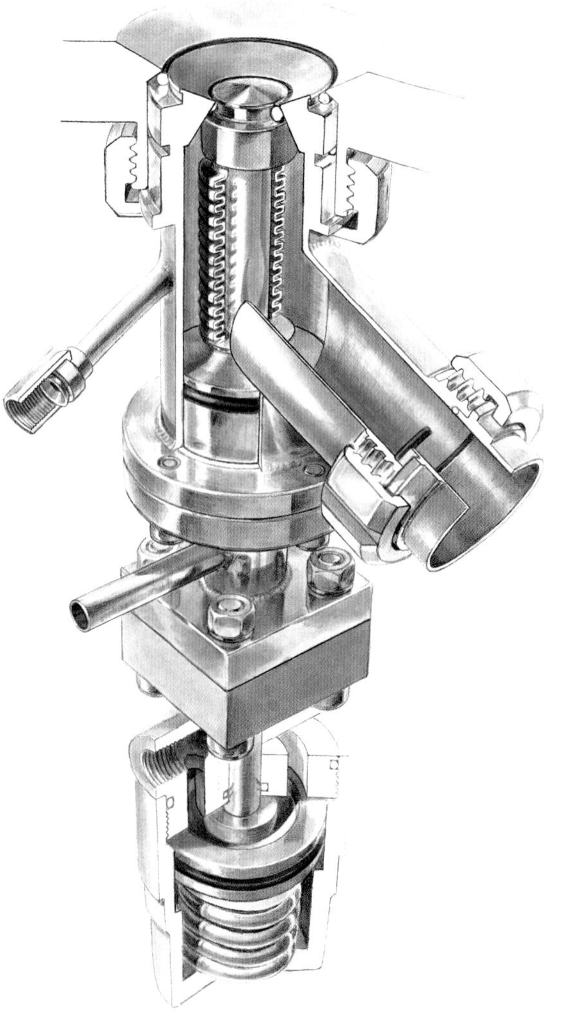

a Welded threaded port
b Sealing head
c Valve stem
d Exhaust
e Drive (pneumatically or manually operated)
f CIP / SIP connection
g Leakage connection

Bottom drain valves are required for draining the reactor. It is very important that the dimensions of the drain be large enough for fast drainage. There is not always sufficient space for adequate fittings. The bellows-type Bioengineering bottom drain valve shown in the diagram is particularly well suited for such applications and has many advantages: large drain cross section on a small area, not permanently welded and therefore flexible with regard to position, not critical with regard to vessel manufacture (welded threaded port), not critical with regard to surfaces to be sealed, no narrowing sections, flexible and easily adaptable, space-saving.

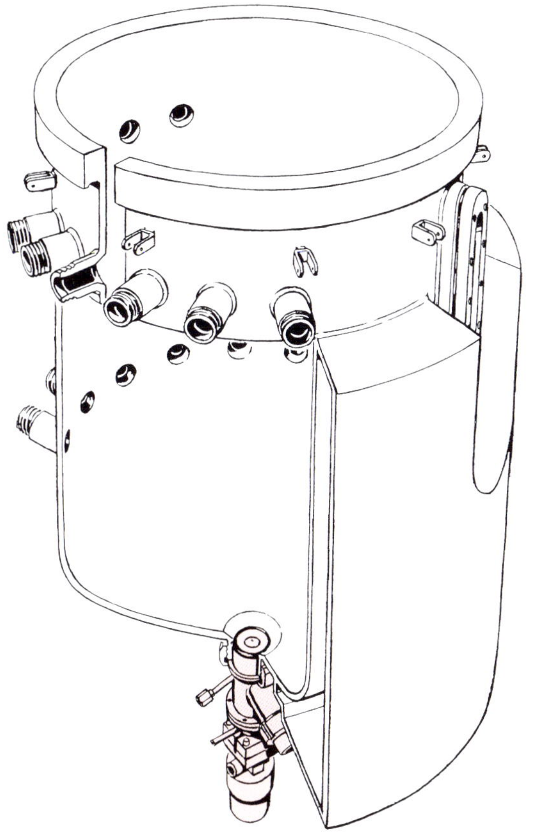

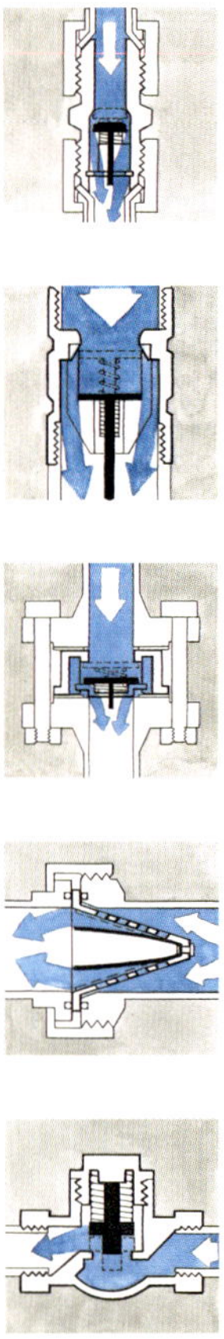

Wide variety of designs. For various media (water, steam, air), vertically or horizontally mounted.

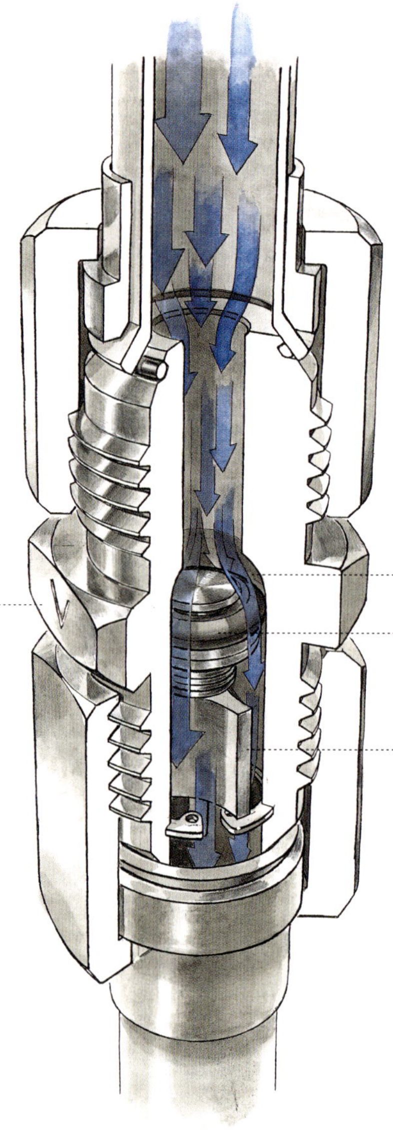

a Medium inlet
b Sealing surface
c Sealing head
d Poppet with spring
e Body
f Medium outlet

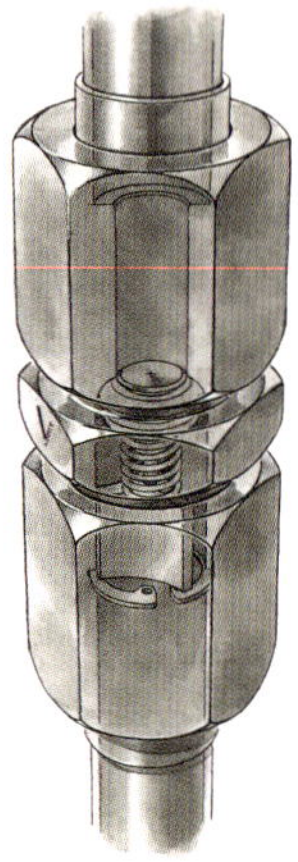

View of a non-return valve in a DN10 line.

Non-return valves are fittings that prevent reflux by a rebound mechanism. The path is opened by the flow. In case of congestion, the valve is closed by the weight of the mechanism, by a spring return, or by the reflux of the medium itself. Bioengineering offers a wide range of non-return valves in many different nominal diameters – designed on the basis of ideas gathered through many years of experience. Examples include fittings in sterile piping for clear segregation of the media.

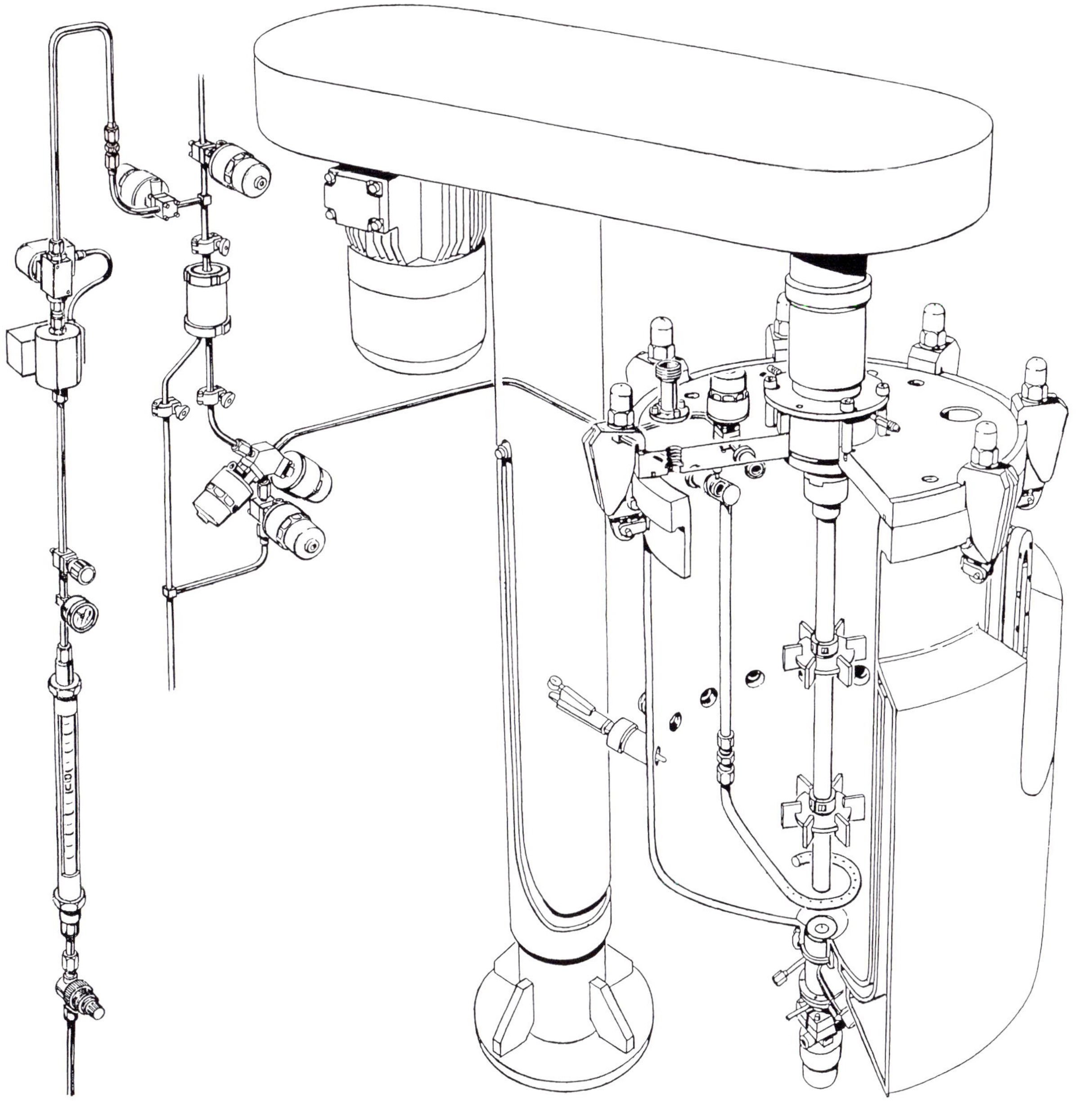

Non-return valves are used wherever a reflux of the flowing medium must be avoided.

Closed Open

Thermostatic capsule steam trap and steam vent. The self-centering ball valve adjusts itself immediately to changing operating conditions. Capsule steam traps ventilate automatically and drain condensate from all process areas with uniformly small supercooling.

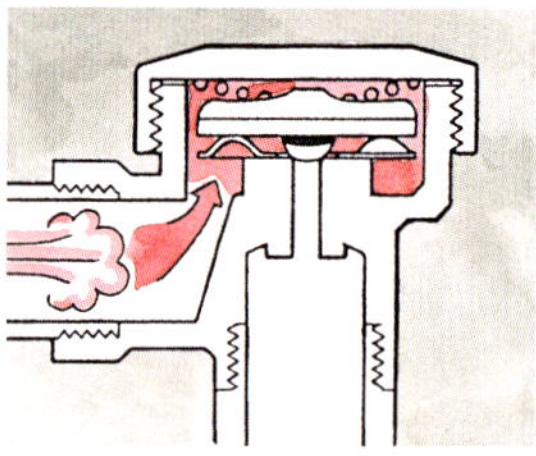

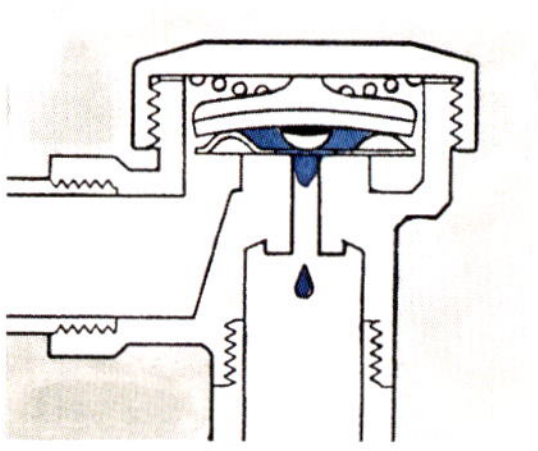

Thermostatic capsule steam trap for clean-steam technology with media-exposed components of stainless steel (316L). Thanks to self-draining design, the traps adjust themselves immediately to changing operating conditions, ventilate automatically, and drain the condensate from all process areas with slight supercooling.

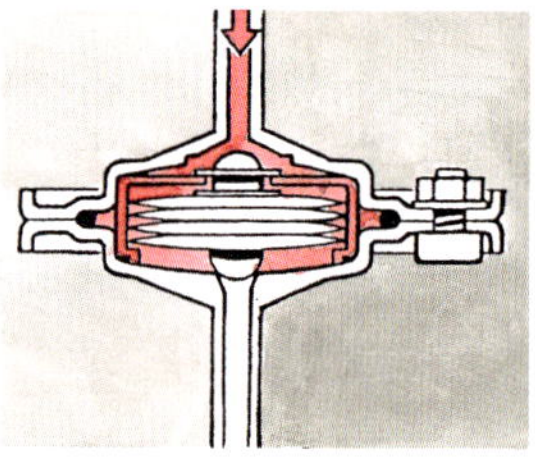

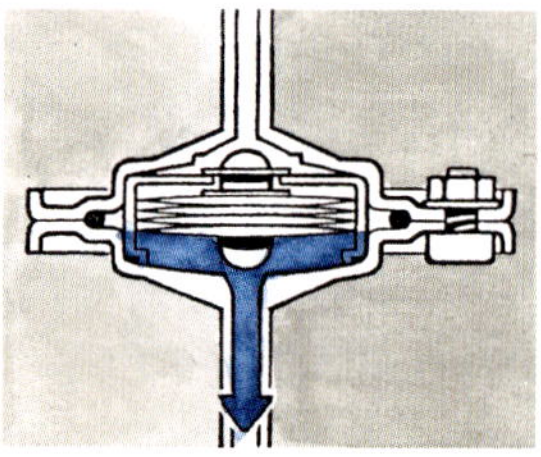

Inverted bucket steam trap. The inverted bucket steam trap operates on the inverted bucket principle and is therefore mounted vertically.

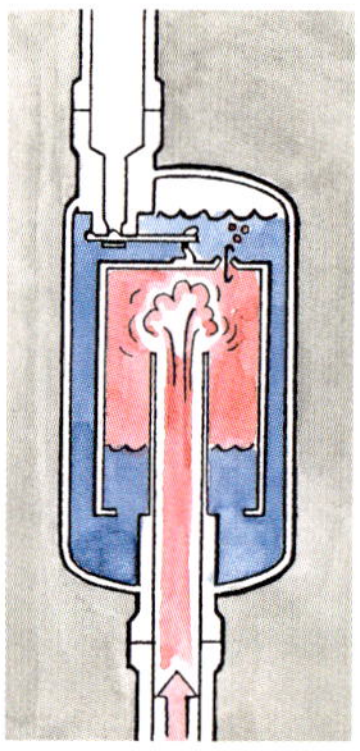

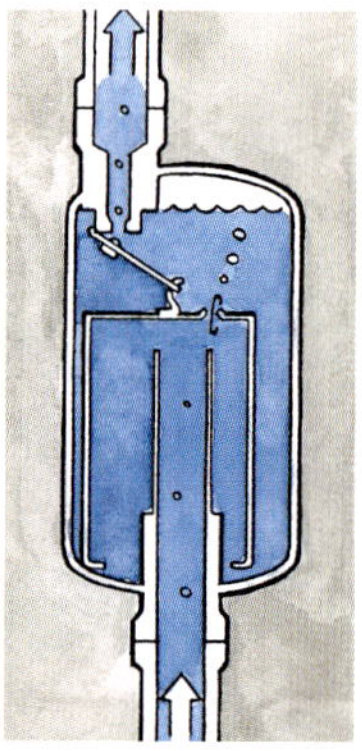

The steam trap is a device that automatically allows condensate, but not steam, to flow through it. The first traps of this kind used inverted buckets and were also called condensate return systems. In the context of Sterilizing in Place (SIP), the steam trap must also double as a temperature controller. Many steam traps cannot handle this function. Bioengineering solves the problem with a simple temperature controller in the sterile sections.

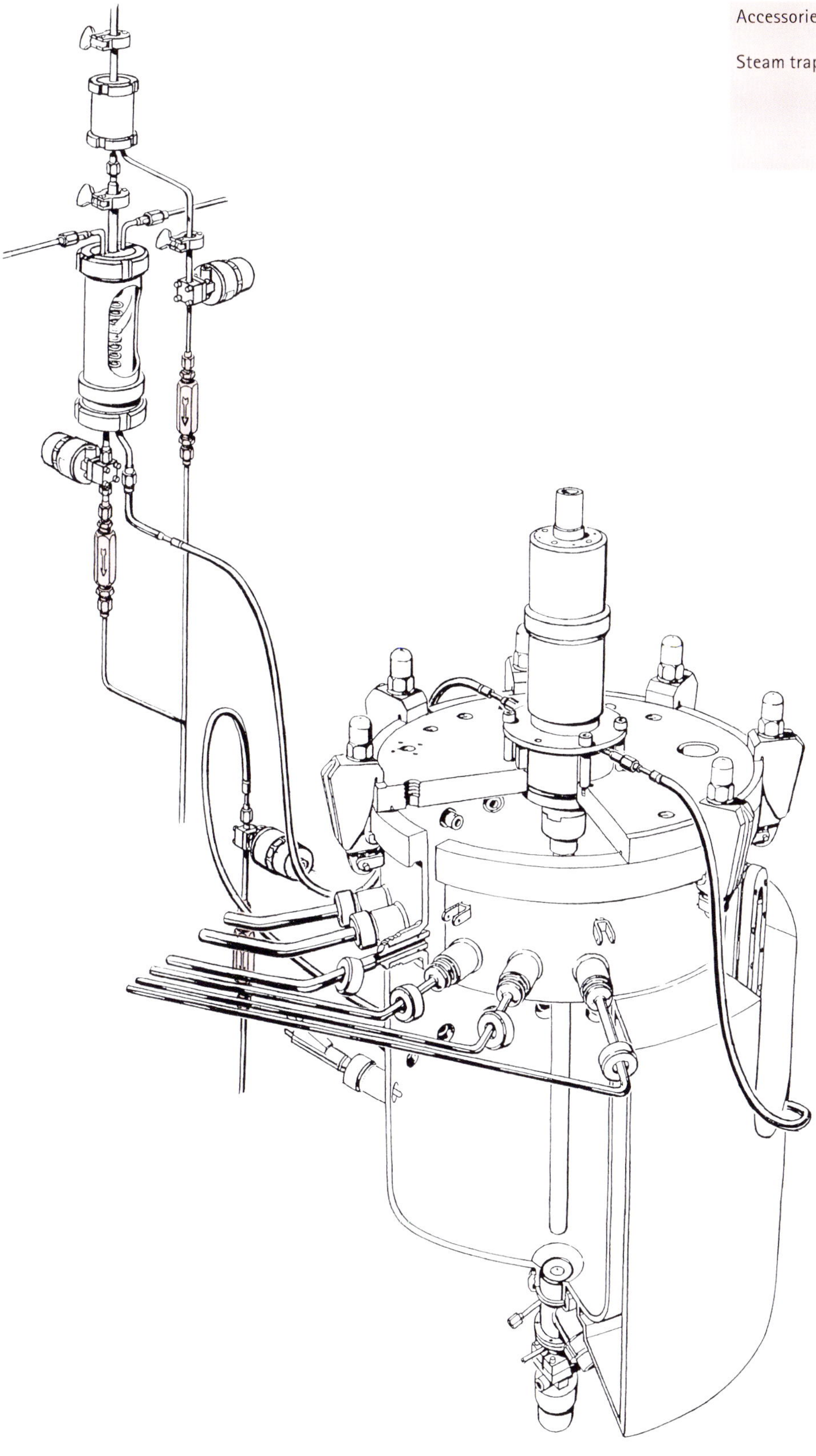

Steam traps are used wherever steam condenses. In the example above, the steam trap is in the balancing section that supplies condensate as a barrier fluid to a mechanical seal.

CupHermann is filled in the airlock, the bag is sealed with a cable tie...

...and connected to the reactor lid via the adapter.

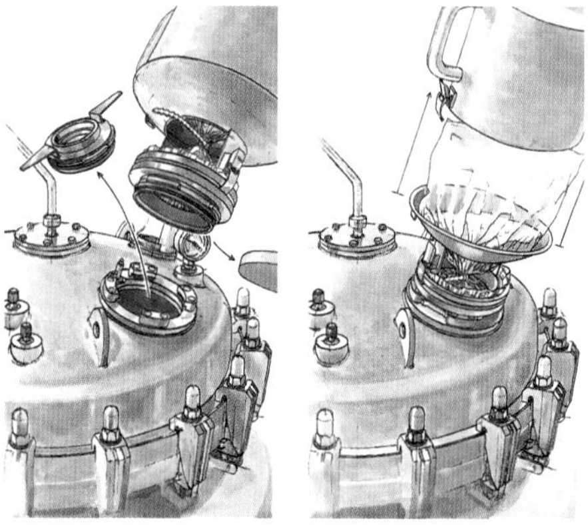

After removing the cable tie, the hopper vessel can be opened and the bag becomes accessible. The residual dust particles can be completely shaken off or rinsed off if necessary.

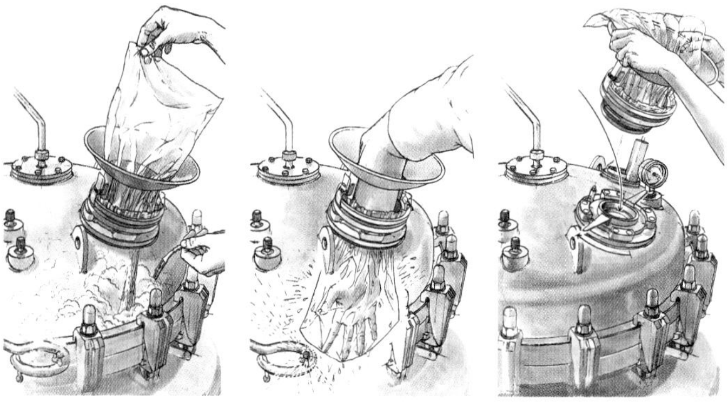

The CupHermann powder handling system developed by Bioengineering is a comprehensive solution for dust-free powder handling from weighbatching to feeding into the agent-processing reactor. Because no dust is suctioned into or out of this absolutely closed system, even the smallest amounts of powdered agents can be transported and completely filled without any risk of contamination. Occupational safety is ensured from contamination-free operation to the disposal of the antistatic bag.

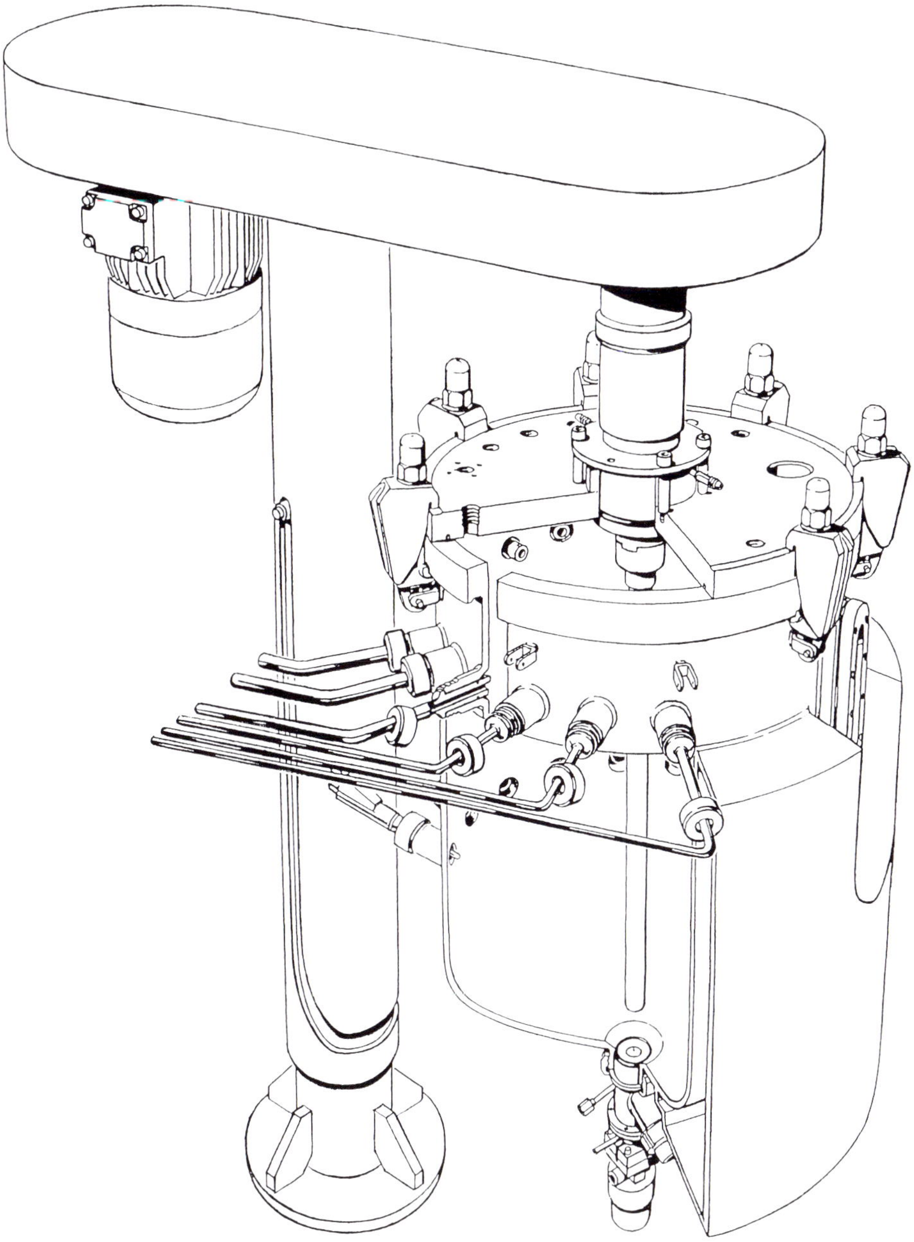

Bioengineering lid ports are suitable for adaptation or retrofitting of various applications such as powder feed, mechanical foam separators, rotor filters, etc.

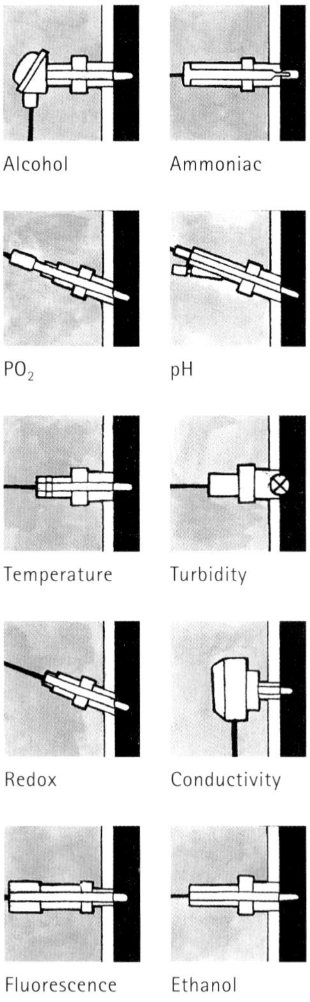

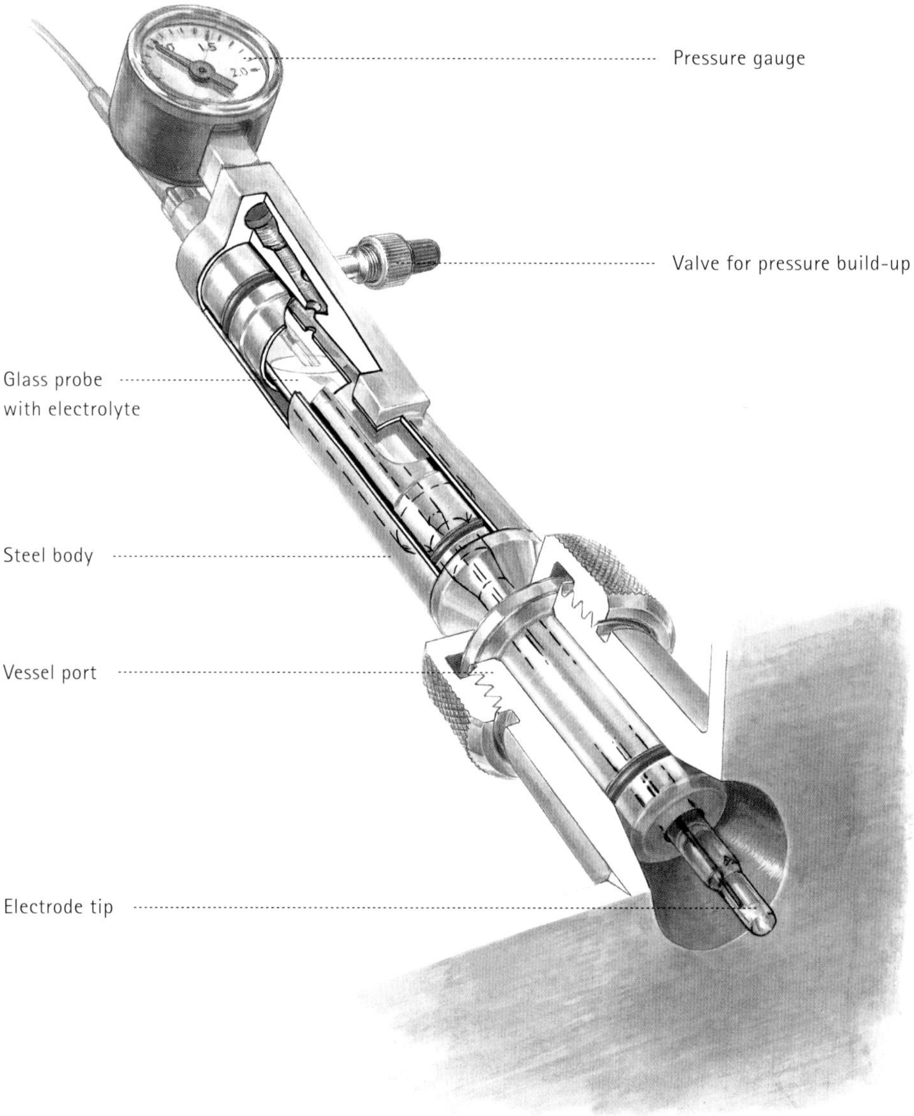

Sensors and probes are used to monitor processes in the reactor. To enable the measurement and control of the smallest possible volumes, all medium-related process parameters such as temperature, oxygen, pH value, turbidity, conductance, etc. are located in the lower part of the reactor. Additional physical parameters such as pressure, level, etc., can normally be read on the upper portion of the reactor. For the measurement of these parameters, many components based on various measuring methods are available on the market. The appropriate sensor is selected on the basis of operating conditions, accuracy classes, and constructive design.

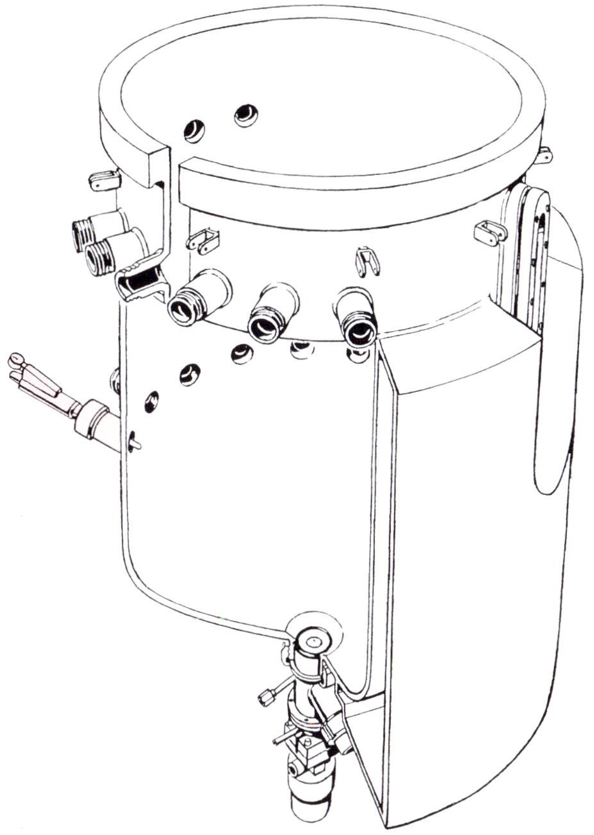

Typical arrangement of a pH sensor near the bottom of the reactor.

Heating loops or circuits are used to cool or heat a medium. They provide a way to control cultivation temperatures or execute sterilization processes. Bioengineering offers a wide range of heating circuit designs adaptable to such tasks. The main elements of a heating circuit include: Safety device to prevent exceeding the permissible service pressure, recirculation pump for recirculating the cooling / heating agent, and at least one power component for heating. Power components can be electrical or hot-water / steam heat exchangers. The selection of these components depends on the energy sources available on site as well as on the size of the unit to be heated or cooled.

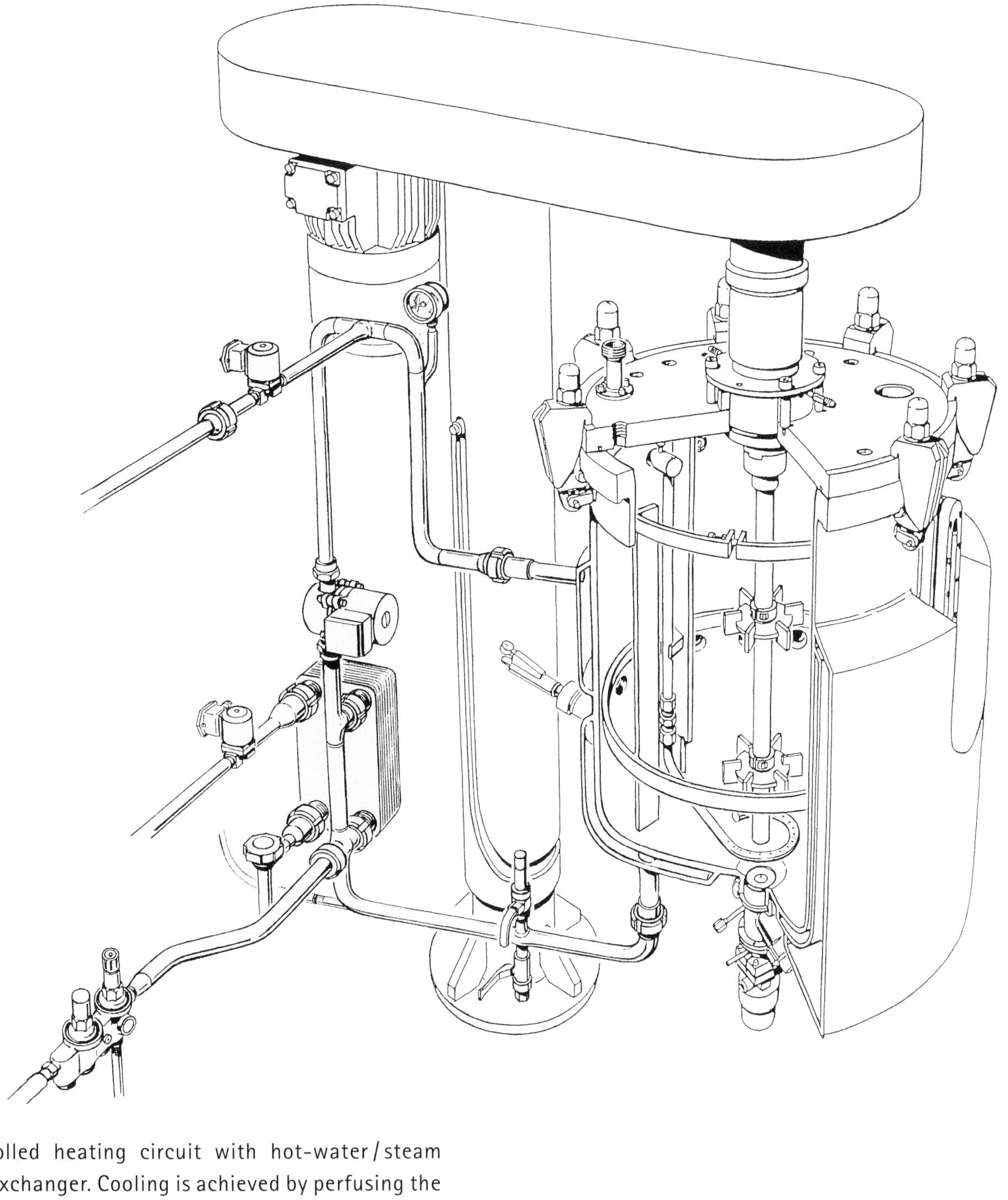

Controlled heating circuit with hot-water/steam heat exchanger. Cooling is achieved by perfusing the circuit with cold fresh water while simultaneously letting off hot circuit water until the desired temperature is attained.

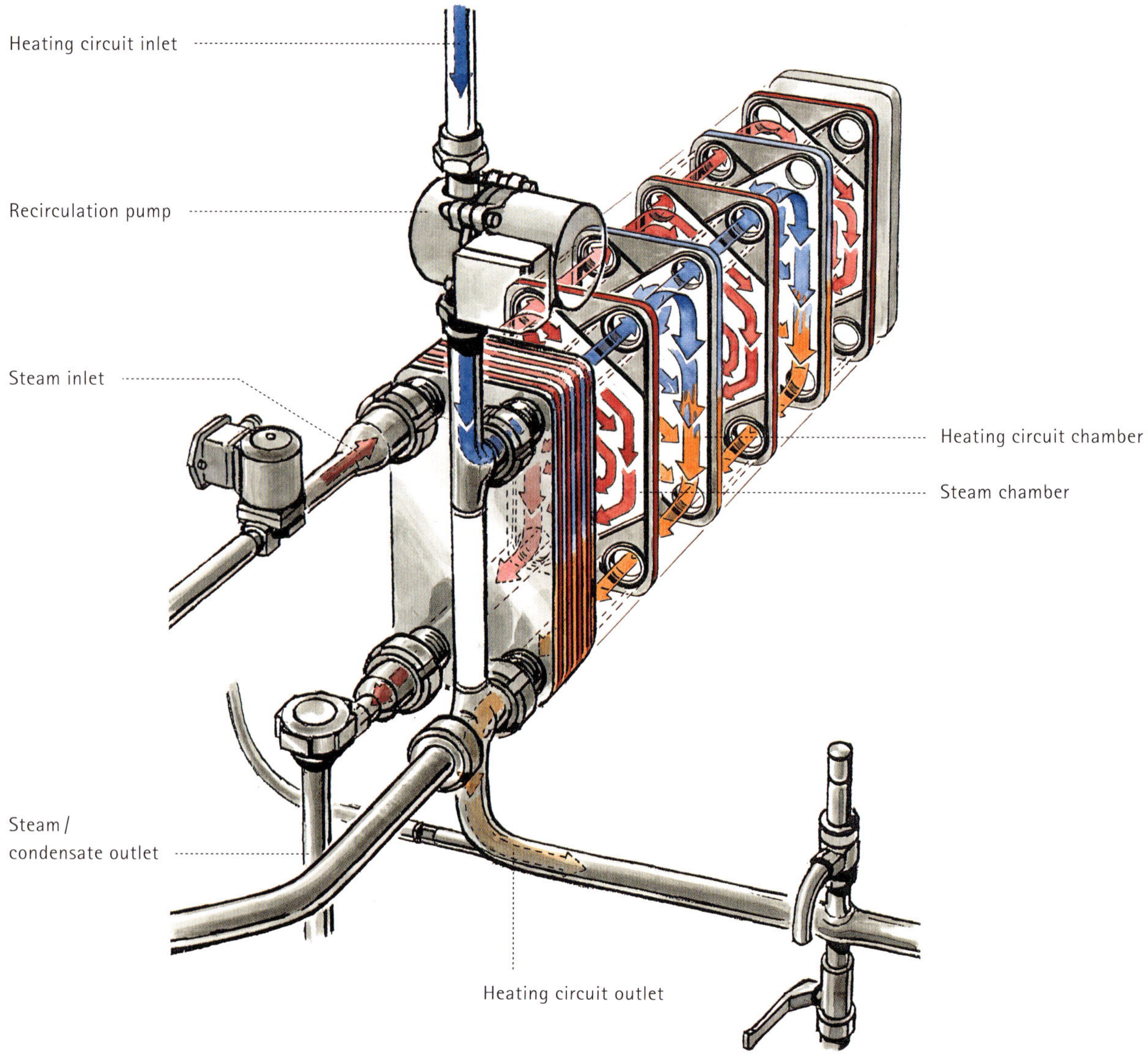

Plate heat exchangers are used as heat-exchange elements in heating circuits. The heating circuit medium and the steam are routed through two separate adjacent chambers using the counterflow principle. The very thin-walled and specially shaped exchangers generate turbulent flow and thus expedite the transfer of heat. After a certain operating time, the efficiency of a heat exchanger may continuously deteriorate as a result of deposits on the surface (e.g. due to hard water). To remove depositions, it is necessary to rinse the heat exchangers periodically with rinsing agents containing acid solutions (e.g. phosphoric acid).

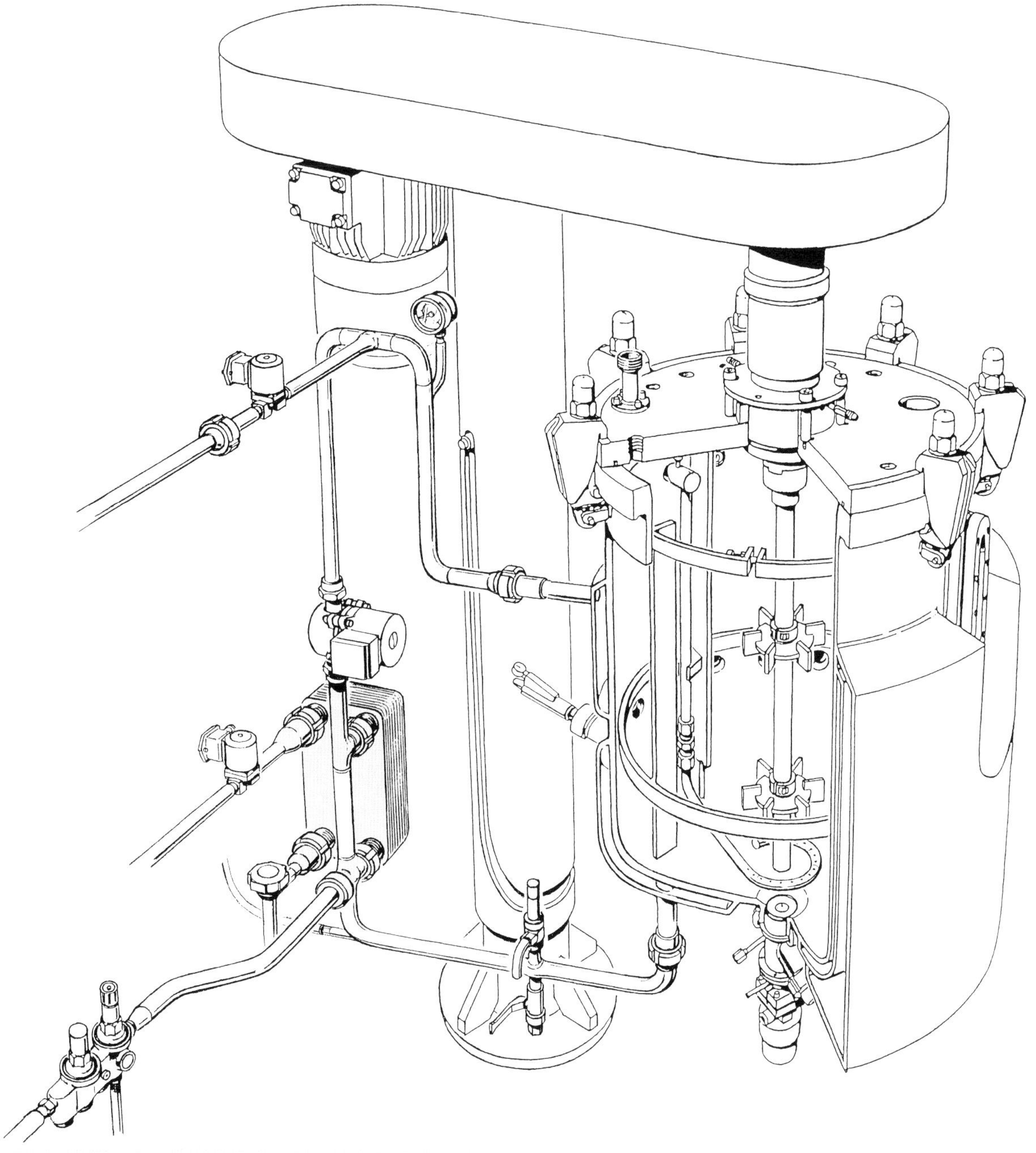

Plate heat exchanger in a heating circuit.

And finally, there it is. The fermenter. A paradigm of superb craftsmanship, equipped with top-quality components, custom-designed for the intended processes. It is the result of technical analysis, careful planning, circumspect procurement, design implementations, manufacturing processes, quality-assurance measures, documentation, and factory acceptance tests. With the exclusive aim of building excellent products for sophisticated users.

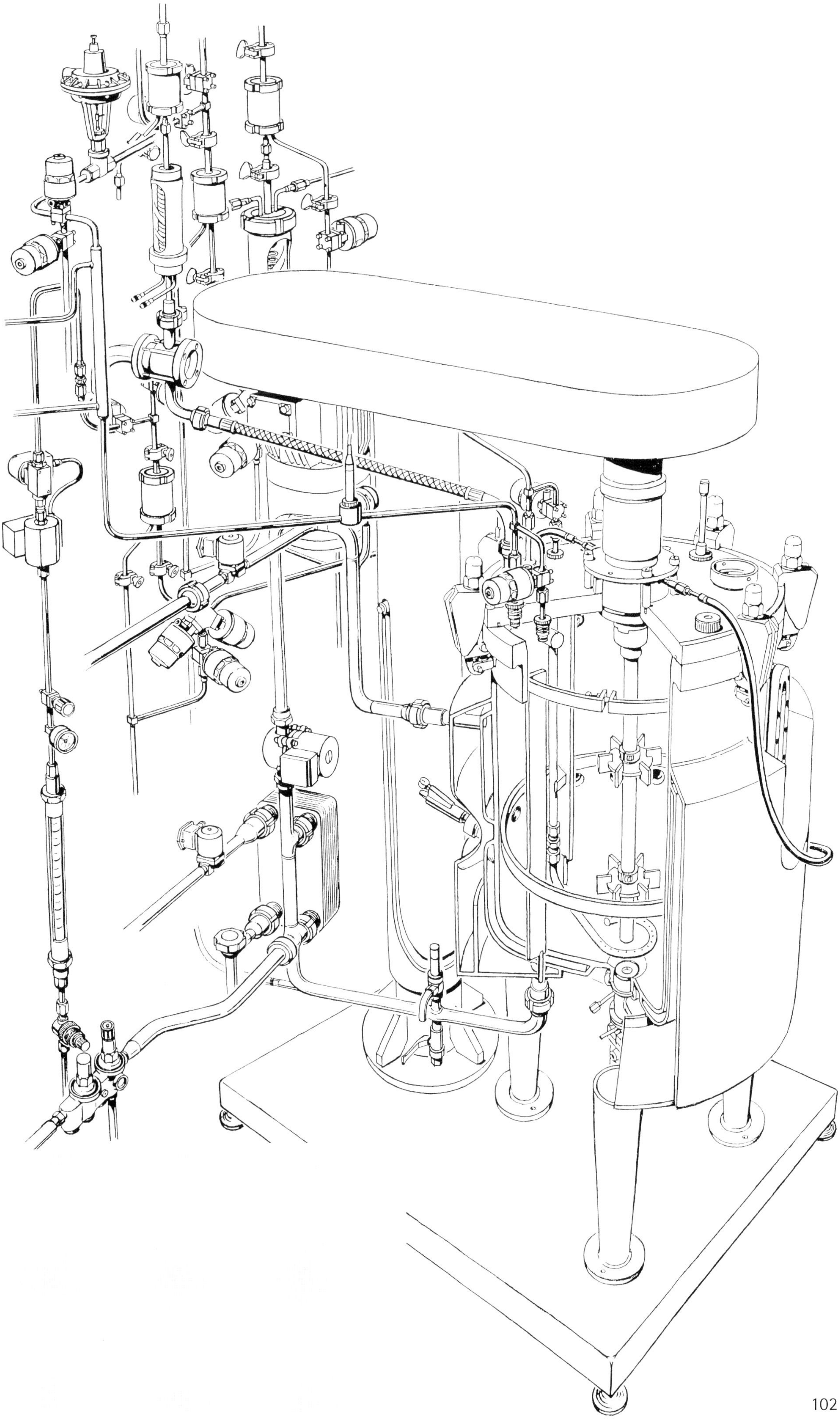